# Advanced Fault Location in Power Transmission Lines



By
Lokesh Yadav

# Contents

# Chapter 1 Introduction

## 1.1 Introduction

The basic functions of our modern society are dependent on electricity supplies which are heavily affected by interruptions. Power system blackouts result in complete interruption of electricity supply to all consumers in a large area. The risk of large blackouts is no longer acceptable and can lead to very large and unexpected social and financial consequences. Reduction of the risk of large system-wide disturbances and blackouts requires that system protection system be approached with the assistance of modern technologies in support of preserving system integrity under adverse conditions. Relaying technology has developed and changed rapidly over the last few decades but the principle objective has remained the same; that is to detect the fault using mostly local voltage and current. In recent years, communication technology has developed very fast which allows distributed protection and control functions in power system. Furthermore, fast and reliable data communication is cheap and widely available. Therefore, new protective functions can be developed in the electrical system based on distributed measured values which are available with fast and reliable communication. IEC 61850 standardizes the communication network and systems for power utility automation which allows free exchange of information between

Intelligent Electronic Devices (IEDs). IEC 61850 enables the integration of protection, automation and monitoring functions which can introduce new opportunity to enhance the performance of the protection and control functions [1].

To improve the reliability of transmission system, this research focuses on transmission line protection and control functions by the implementation of the IEC 61850.

## 1.2 Basic Definition

*Intelligent Electronic Devices (IEDs):* IED is a microprocessor based device that performs various automation function e.g. protection, control, and measurement.

*Interoperability:* the ability of dissimilar information technology system to talk effectively without the intervention of users. In the other words, it is the ability of two or more devices of different vender to exchange information.

*Protocol:* A set of rules to communicate data with another device. The rules define syntax, structure, and format of data.

*Fibre Optic Current Sensor:* is a sensor which measures the current of line by utilizing of the magneto-optic effect (Faraday Effect).

*GOOSE:* is a high priority message which is used for fast communication in SAS.

*Protection system:* A system which protects equipment from damage due to fault in power system.

*Fault Location:* An estimation of the distance to the fault in a transmission line.

*COMTRADE: A* format for transient data exchange which is used for analysis of undesirable sources faults, harmonics and analysis of events which causing power system instability.

*LAN:* A network which is built within a small area. It does not need the Internet connection to transmit data between different devices.

*WAN*: A network which is the connections between LANs or other networks. The data can be transmitted over longer distance.

*OSI model*: The OSI (Open System Interconnection) model is a reference on how to build up the messages to be sent in a network. Based on this model, the data needs to be sent through the seven layers where each layer modifies the data as the protocol rules. The process is reversed in the received location.

*Protective relay:* It is a device for supervising and protecting the electrical system from any kind of faults.

*Merging Unit*: The interface unit that convert measured values inputs/outputs to digital format based on IEC 61850 standards.

*Sampled Value*: IEC 61850-9-2 is used to transmit the measured voltage and current signals by Instrument Transformers. The digitally formatted and time stamped multicast Sampled Values are transmitted via Fibre Optic to IEDs.

*Logical Node (LN)*: LNs represent functions or equipment which are used in the power system based on IEC 61850 standards [2].

*Substation Configuration description Language (SCL)*: It allows the description of an IED configuration in communication and engineering applications. The main purpose of this part is to provide interoperability between an IED and system configuration tools from various suppliers [3].

*Reliability:* The inverse of the probability of failure is quantitative reliability, where less failure is better reliability. Also, fast fault clearance can increase the reliability.

## 1.3 Main Components of Transmission Line Protection Systems

More than 85% of power system faults have occurred in transmission lines. The objective of each protection function is to isolate faults with maximum selectivity and as fast as possible. Maximum selectivity means that only the breakers nearest to the fault are tripped by the protection relays. There are different line protection functions. However, the distance and differential protection functions are better and they can provide selective tripping. Following subsections have been discussed the main components in the transmission line protection system.

### 1.3.1 Transmission Line

A transmission line can be overhead, underground, or combination of them. Because of air is used as the insulation medium between phases, overhead lines are more economical than underground cables for long distance. Overhead transmission lines are prone to nature phenomena such as ice storms, lightning strokes, and etc. Whereas underground transmission line is better shielded against external hazards when a fault happens the localization of the fault is slower than with overhead transmission line. Because of high resistance and reactance involved with long transmission lines, the protection of this kind of lines is more demanding than the of short ones.

### 1.3.2 Substation

Substations are the nodes of the transmission system, where circuit breakers, transformers, and substation automation devices are located with the ability to reconfigure the system topology, isolate equipment for repair, control, monitor and communicate to the control centre, protective relaying decisions, etc. One of the fastest evolving parts of the modern electrical system is Substation Automation System (SAS) which is the most flexible part for developing new protection and control functions.

### 1.3.3 Tele-communication

The line protection system should be equipped with communication when a fast trip is essential. Distance and differential relays use telecommunication channel. The communication channel can be power line carrier (PLC), optic fibre, and radio. To increase the reliability of line protection system, a duplicated communication channel can be applied. In this case, these two channels should have the separate route and power supply.

## 1.4 Communication in Protective Relays

The protection relays have been seen as a vital part of any substations which it does not need to communicate outside their protection zones. However, this has been changed due to development of communication technology. Previously the Relay communications were to facilitate the programming, change settings, installation, and maintenance, without any thought of real time communications. The first communications to relay were based on RS-232 and RS-484. Each manufacturer developed its own protocol. Today, IEDs provide protections, controls, and monitoring function together with communications.

## 1.5 Fiber-Optic Current and Voltage Sensors for High-Voltage Substations

Conventional instrument transformers utilize the magnetic field to measure the voltage and current. The concept of "smart instrument" that have local intelligence and advanced communication capabilities, has influenced the electrical protection system industry. They have been developed which convert the analog measured values into digital, to be sent to the protection relays over a communication channel. Although data communication does not commonly apply to the process bus as yet, the very near future will see data communication being utilized on this level more and more.

The newly developed fiber-optic transducer which uses the Faraday 's effect in the fiber, open new perspective in the field of measurement in power system. The Faraday 's effect is the polarization of light waves by the magnetic fields. Fiber-optic instrument transformers are adapted to high-voltage environments as they are immune to electro-magnetic interference. The wide bandwidth of optical transducers is important for fast protection. They can be easily installed in significant space and cost saving. Figure 1-1 shows a FOCS developed by ABB for high voltage substation. The reduced size and weight are attractive benefits compare to conventional ones. Unlike traditional instrument transformer, it does not need to be a standalone device. Instead, it can be integrated into the primary high voltage equipment. Another important feature of this new technology is its ability to communicate via optical process bus.

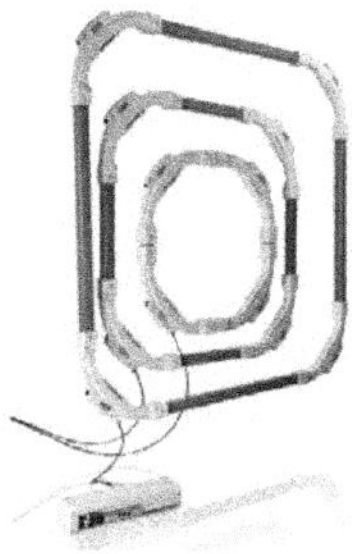

Figure 1-1- Fiber Optic Current Sensors

In this thesis, new transmission line protection functions are evaluated based on distributed measured values by fiber optic instrument transformers which are installed in the different location of a line.

## 1.6 IEC 61850 Standard

IEC 61850 is a communication standard series for power substations developed by International Electro-Technical Commission in the year 2000. The title of the Standard is: "Communication networks and systems for power utility automation". The title of the standard should not be misinterpreted as a communication protocol which may cause to underestimate the benefits of the standard. Generally, communication protocols deal with the entire structure of the message frame in order to convey a message. However, the IEC 61850 provides configuration of IEDs with interoperability to communication using a protocol. It provides the engineering definitions for configuration of the functions required for communication between IEDs, and the related system requirements. Hence, it is more than a communication protocol.

### 1.6.1 Concepts

IEC 61850 integrate all automation functions within a substation. The function can be split into LNs which are placed in different IEDs and communicate with each other. The IEC 61850 standard uses the concept of virtualization to provide a view of a real device for exchanging the information between devices. Figure 1-2 illustrates the approach model of real devices in a substation [4].

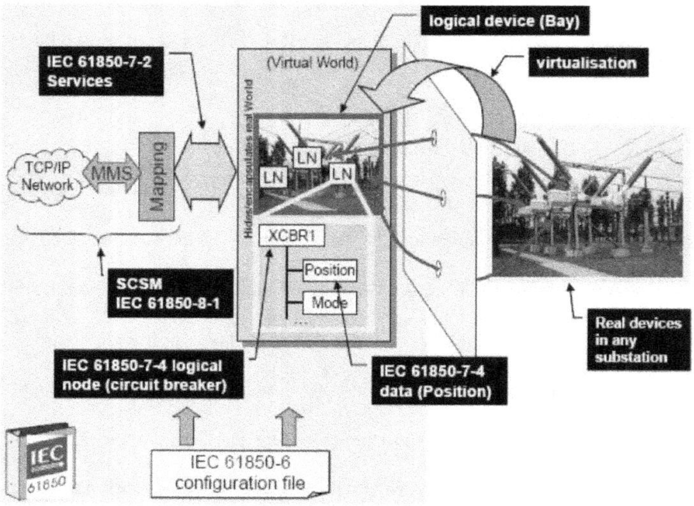


Figure 1-2- Modelling approach (conceptual) [4]

Only the information and exchange information which is required to provide interoperability of devices is defined in IEC 61850.

### 1.6.2 The Data Model

The data model is divided into Logical Device, Logical Node, Data Classes, and data attributes. An IED can house multiple Logical Device and Logical Node in it. Also, each LN is a collection of data classes where the possible values of data classes are called data attributes. Figure 1-3 shows the data model layers in IEC 61850.

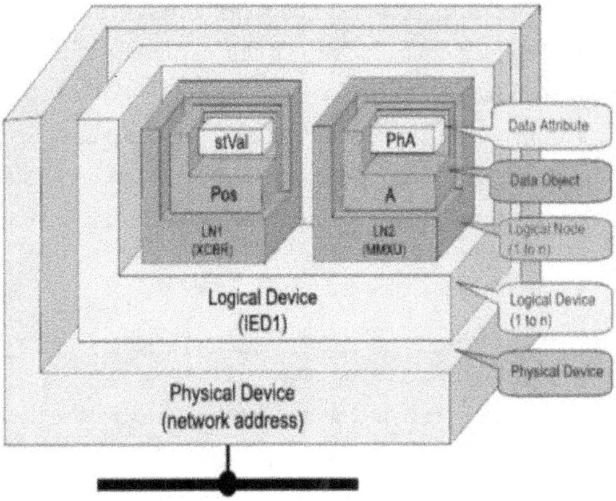


Figure 1-3- Data model in IEC 61850

### 1.6.3 Topology and Communication Interfaces of SAS

There are three hierarchical levels (station, bay, or process) of a SAS based on IEC 61850. The IEC 61850 edition I standardize the communication within each level and between different levels. However, the requirement to exchange data between substations and between substations and control centre is increasing due to improve power system operation and protection applications. The second edition of the standard provides the basic features to be used for that data exchange. The functions of a SAS may be allocated logically on the different levels. Figure 1-4 shows the communication interfaces of these levels which are illustrated by numbers. The numbers refer to the communication interfaces in a SAS as follows:

- Interface 1: protection-data exchange between bay and station level

- Interface 2: protection data exchange between substations
- Interface 3: data exchange within bay level
- Interface 4: SV exchange for CTs and VTs
- Interface 5: control data exchange between process and bay level
- Interface 6: control-data exchange between bay and station level
- Interface 7: data exchange between substation and a remote workplace for engineering and configuration
- Interface 8: fast exchange of I/O for P&C
- Interface 9: data exchange within station level
- Interface 10: communication to control centre
- Interface 11: control data exchange between substations

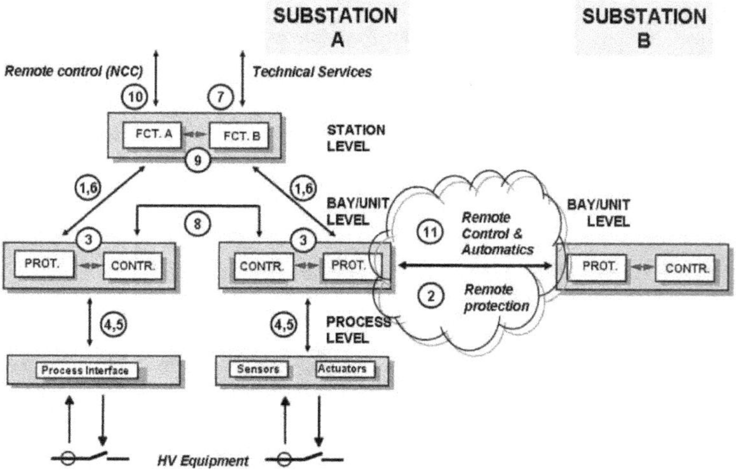


Figure 1-4- Communication architecture in SAS based on the second edition of IEC 61850

### 1.6.4 Abstract Communication Service Interface

IEC 61850 models power system devices using object oriented modeling approach in order to achieve data organization and object configuration. Figure 1-5 presents the mapping of various IEC 61850 services over OSI model. This modeling is mapped onto specific protocols (such as Ethernet) which provide consistency and interoperability. The Sampled Value (SV) and Generic Object Oriented Substation Event (GOOSE) are mapped onto the Ethernet layer of the OSI model in the IEC 61850 standard. The mapped service can save the time in the communication system. Hence it is very useful to apply the service in the time critical application and real time system.

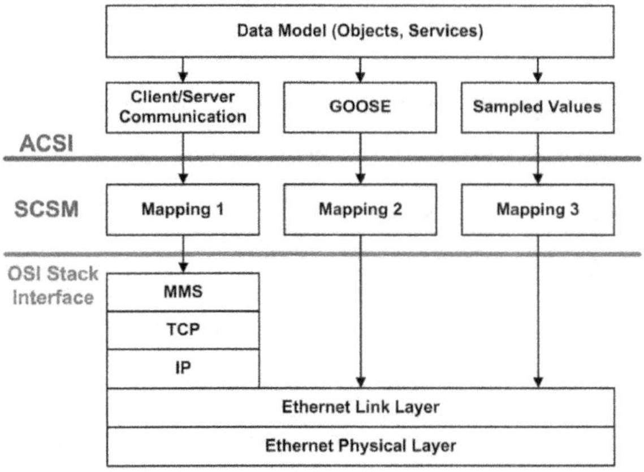


Figure 1-5- IEC 61850 service mapping over ISO/OSI seven-layer model

There are different types of messages which are defined in IEC 61850-5. The Type 1 and Type 1A are GOOSE messages which are time critical in nature such as Trip, Close, Start, Stop etc. They are mapped directly onto Ethernet layer to reduce the stack. The transmission of raw data or Type 4 messages (SVs), which are continuous streams of synchronized data from digital instrument transformers, are also mapped on directly to Ethernet layer with broadcast/multicast addressing. Type 2, 3, and 5 messages corresponds to ASCI applications such as normal 'state' information messages, auto-control functions, transmission of event records, general presentation of system data which are time-tagged messages, file transfer messages etc, which require message oriented services for which MMS protocols with TCP/IP stack above Ethernet layer are used.

## 1.7 Impact of IEC 61850 on Reliability

Reliability in augmentation of the system or replacement of IEDs is perhaps the most profound risk reduction in substations due to the use of IEC 61850. Wire based engineering and systems are prone to errors in installation, construction, fabrication, testing and engineering are the most common causes of faults, blackouts, equipment damage, and personal injury. On the other hand, IEC 61850 bay configuration can be copied to a new bay with a few changes. Such this configuration reduces commissioning time and at the same time human errors. Furthermore, the implementation of IEC 61850-9-2 known as process bus eliminates the cabling to the secondary system IEDs. Therefore, the risk of open CT circuited is completely eliminated with the replacement of Merging Unit and IEC 61850-9-2 Sample Values.

Reliability is partially an outcome of high availability systems. Availability is the ratio of system "up-time" to the total time in service and is given by the formula:

$$A = \frac{MTBF}{MTBF + MTTR} \qquad \text{Eq. 1.1}$$

Where A is the availability of a system, MTTBF is Mean Time To Failure, and MTTR is Mean Time To Restore. Therefore, the reliability or availability is improved by designing out potential failure modes. Also, the restoration time is faster by designing mechanisms. The reliability of IEC 61850 systems is increased due to the ability of continually monitored of the system.

Ultimately the best solution is the possibility to restore the system to service remotely with IEC 61850. Therefore, the reliability and availability of system can be significantly increased by replacing of hard wire with network and IEC 61850. However, the reliability of the network is very important in this system. The reliability of different network topologies is evaluated in this thesis.

## 1.8 Objectives

The aim of this research work is to improve the reliability of transmission line protection system with the communication technology. Following are the main objectives of this research:

- To develop a laboratory setup to do the experimental study in transmission line fault location and protection
- To evaluate existing technique for protecting and operating of a transmission line based on communication technology
- To integrate TW-based fault location into the existing protection relays
- To evaluate IEC 61850 based accelerate distance protection
- To evaluate performance of IEC 61850 GOOSE in WAN
- To evaluate reliability of IEC 61850 based substation automation system and process bus

## 1.9 Organization of the Thesis

This thesis is organized into seven chapters to provide appropriate emphasis on the topics related to the work done. The chapters of this thesis are organized as follows:

Chapter 2 provides the review of the available research work in the areas of research done in this thesis. Chapter 3 propose a new method in impedance-based and travelling wave- based fault location. Where in impedance-based fault location, distributed measured values of a line are applied, and in travelling wave method, detection of the ground fault with measured zero sequences is proposed. Chapter 4 discusses the transmission line protection functions. The performance of distance protection and differential protection are compared in PSCAD software. A new differential protection function is developed for the measured values in a line. The feasibility of the proposed scheme is tested in PSCAD and its performance is compared with the existing differential protection function. Chapter 5 discusses substation automation based on IEC 61850. IEC 61850 GOOSE based Interlocking and accelerated distance protection are introduced in this chapter. The practical consideration of IEC 61850 GOOSE and the advantages of the scheme are reported in this chapter. Chapter 6 discusses the assessment of reliability and availability in the different structure of SAS to increase the reliability of the feeder protection systems. Conclusions and the suggestions for the future work are discussed in the last chapter.

## 1.10 Conclusion

The requirement for communication based transmission line protection system has been discussed in this chapter. An introduction to the transmission line protection components is presented. New instrument transformer with fiber optic technology is introduced to measure the voltage and current. The main scope and features of IEC 61850, the communication standard for the power system, are presented in this chapter.

The next chapter is an extensive review of literature related to research work to justify the research work carried out in this thesis.

# Chapter 2 Literature Review

## 2.1 Introduction

To gain the right perspective of the research problem, an extensive literature review has been carried out in the related areas of the research work done. This chapter presents a brief review of the literature in the related areas of research work done in this thesis. The literature review is categorized in the following broad categories.

- Fault location in transmission line
    - Impedance-based fault location
    - Travelling wave-based fault location
- Transmission line protection functions
- IEC 61850 based Substation Automation System (SAS)
- Reliability and availability assessment of SAS

## 2.2 Fault Location in Transmission Line

Accurate fault location can expedite the repair of the faulted components, speed-up restoration, reduce outage time. Therefore, it can improve power system reliability [5, 6]. A review of existing fault location methods is conducted to identify the most suited fault location methods in Transmission Line (TL). The current fault location techniques can be divided into impedance-and travelling wave-based fault location methods. Following subsections review the impedance- and TW-based fault location methods.

### 2.2.1 Impendence based Fault Location

Impedance-based fault location methods are the most used by utilities [7, 8]. The impedance-based fault location technique compares the line parameters to the impedance measured at the fault time. The distance estimation to the fault can be obtained with the comparison of these values. Many algorithms have been developed to improve the accuracy of this technique. The current impedance-based fault location algorithm can be categorized as:

- One-end methods [9, 10]
- Two-end methods [11-13]

One-end impedance fault location methods are economic and simple compare to two-end methods [9]. One end impedance fault locators calculate the location of a fault based on the impedance from one end of a line. This technique is subjected to several sources of error, such as the zero sequence mutual effects, the uncertainty of parameter in transmission line, unbalanced load flow, the influence of FACTS devices, the accuracy of transmission line model, and measurements errors [14].

Two-end algorithms estimate the location of fault using voltage and current from both ends of a line. Based on this technique, transient fault data must be collected and synchronized from both ends of a transmission line. The transient data of faults can be reported by the protection relays or Digital Fault Recorders (DFRs) which are installed in substations. In order to have an accurate fault location, the data from both ends of a transmission line should be analysed [12]. Therefore, in this technique, microprocessor relays (or DFRs), communication facility, and analysis software are the most crucial requirements. Two-end fault location methods become more common with the development of communication technology [15, 16].

Two-end algorithms have two classes. One of them makes use of synchronized data, and the other one utilizes unsynchronized data. In the synchronized class, the measured values should be synchronized by global positioning system (GPS). Researchers [11, 12, 17, 18] have utilized this technique to increase the accuracy of fault location. One the other hand, in unsynchronized class, the GPS system is not required [12, 13]. This technique is more economical and easier to implement. However, the accuracy of the technique is less compare to synchronized method.

There are many factors which can affect fault location accuracy. [19, 20] have considered the influence of the pre fault load. [15, 21] discus CT saturation in fault location. To omit the influence of the fault resistance, some of the early methods only utilise the imaginary part in single-end fault location estimation [22, 23]. However, in double-end, the current from the far end will contribute to the reactance measured by the fault locator [24]. Fault resistance on single end method is the most effective in fault location inaccuracy. Takagi, and etc in [25] have proposed a method to compensate the influence of the fault resistance.

- *Literature review on fault location has been found that great effort has been made in the past to develop various algorithms to improve fault location and protection of transmissions lines, which can be classified into either the one terminal method or the multi-terminal method. One end method only use the measured values of the one side of the line. However, the accuracy of these method may be affected by the remote terminal system's impedance and the fault resistanc. This research puts forward a novel fault-location algorithm to achieve the better accuracy of fault location. The proposed method is discussed in the next chapter.*

### 2.2.2 Travelling Wave Based Fault Location

Travelling Wave (TW) based fault location algorithms have been increasingly investigated due to advancement in the technology, especially digital electronics, satellite based synchronization, and digital communications. In this method of fault location, each terminal of a line calculates the accurate time of the arrived surge due to a fault. As the surge speed is same as the speed of light, comparing the surge wave arrived on each terminal of a line can estimate the location of faults. Using the standard time reference at both the sides is the primary requirement in this technique. Therefore, communication system, accurate time stamping, appropriate current or voltage sensor and TW protection relay are the basic requirements in this method.

At the time instant when a fault occurs, transient waves of voltage and current travel from the fault point towards both ends of the line [26]. Travelling wave based- fault location technique identifies the arrival time of these waves to estimate the distance to the fault. Depending how information extracted from captured ware, there are five types of scenario in TW based fault location.

The first type (Figure 2-1), measure the difference of time between the first arrival and its reflection. In this method, information at one end is required. [27, 28] used this technique to estimate the location of fault in distribution grid with distributed generation and a transmission line.

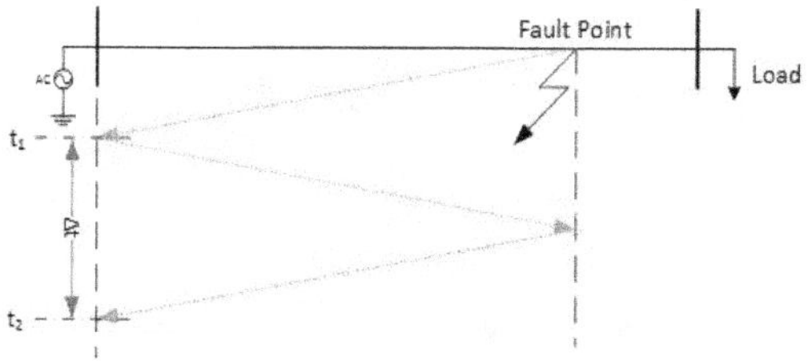


Figure 2-1 - Type one: first arrival and its reflection

In the second type (Figure 2-2), a timer starts counting the time when a travelling wave is received at one terminal until remote end sends the stop signal. As the delay of the communication channel is clear, the total time difference is calculated. Consequently, the distance to the fault is estimated by utilizing time difference and the time of the first arrival. The authors [29] used this technique to estimate fault location on a transmission line.

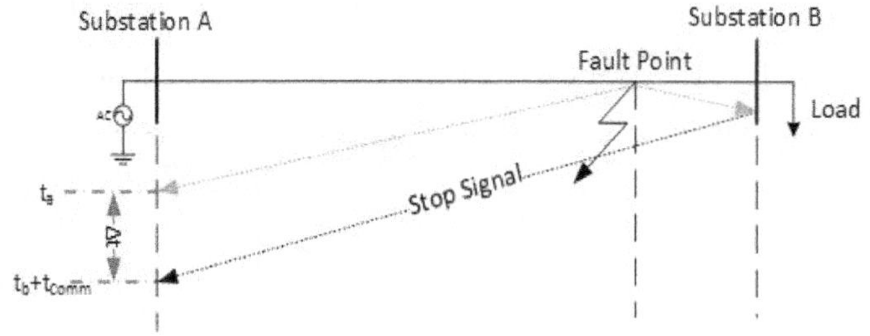


Figure 2-2 - Type two: first arrival and stop signal from remote end

In the third type (Figure 2-3), an impulse signal is injected into a transmission line after fault detection at the relay point. The location of a fault is estimated by measuring the time difference between the fault detection and receiving the reflected wave. This technique utilizes the signal in the local substation and the data from the remote substation is not required.

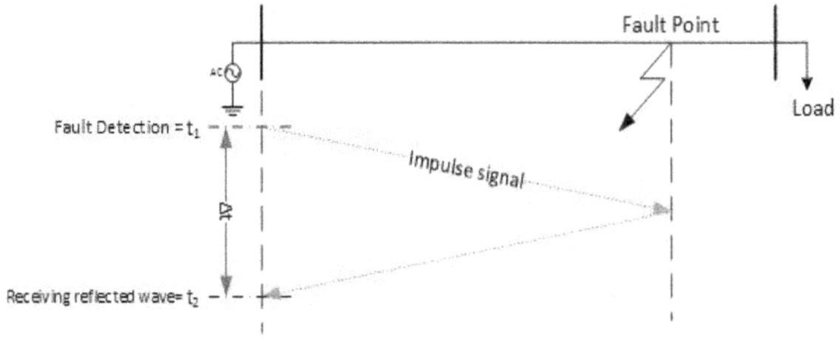


Figure 2-3 - Type three: impulse signal injection from one end

In the fourth type (Figure 2-4), an accurate time stamp is required. When a travelling wave arrives at the both ends of a line, it would be stamped by time. The data from all terminals is sent to a central processing unit to analyse and estimate the fault location.

It is clear that the accuracy of this technique is highly dependent on the accuracy of data sampling. [30] used this technique to estimate fault location on half wavelength transmission line. And [31] used this technique to estimate the fault location for hybrid multi-terminal circuits.

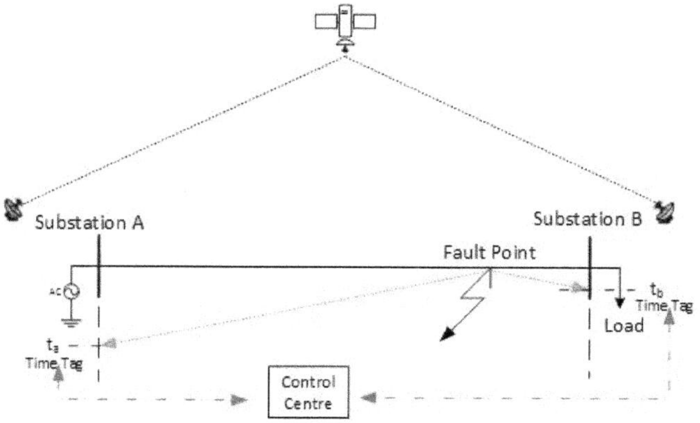


Figure 2-4 - Type four: Time stamp by GPS clock

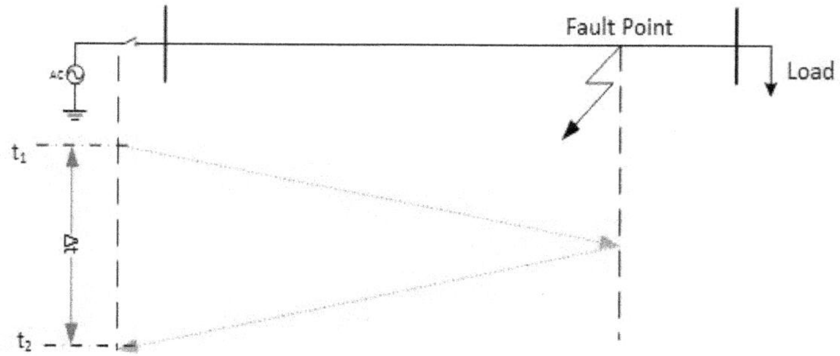


Figure 2-5 - Type five: transient data due to the opening of circuit breaker

In the fifth type (Figure 2-5), the data of generated transient due to the opening of contacts of circuit breaker is used. This is same as generating an impulse wave in the system. [32] used this technique and frequency analysis to find the fault location.

For all of the above-mentioned techniques, the accuracy of TW-based fault location technique depends on the accuracy of fault detection system and the time stamping system. Several papers have been published in which different algorithms have been proposed to improve the accuracy of TW based fault location. The authors [7] propose to filter the power frequency from Current Transformers (CTs) by high-pass filter and sample the remained frequency to determine whether a fault happened on the line or not. This method needs a high sampling frequency of three phase signals. The authors [33] developed a function for TW-based fault location using IEC 61850, and they used transient fault report data in their function. However, in a practical case, the sampling frequency of the IEC 61850-9-2 Sample Values (SV) is usually less than 20 kHz which is not sufficient for TW based fault location.

The wavelet transform is a powerful method to detect fault from the analysis of transient fault data [34-36]. It can be used to extract some features of a signal [37]. Thereafter, many scholars have applied these features in their algorithms, to improve the accuracy of TW based fault location. For instance, combination of wavelet and Artificial Neural Network (ANN) has been applied to estimate the distance to fault [38]; The authors [39] applied Hilbert discrete wavelet transform; [40] used neural network and wavelet transform; The authors [41, 42] proposed TW-based fault location for multi terminal circuits.

- *There has been extensive research on TW based fault location. However, very few of them have considered faulting detection strategy. While many algorithms have been developed in simulation software, the fault detection for TW-based fault location in transmission lines still has technical problems. One of the main problems is that most of the algorithms need a high sampling frequency of three phase system which is costly in application. This thesis proposes the detection of zero sequence components which is a digital value and easy to monitor with high sampling frequency. However, this technique can be applied only on ground faults. On this basis, the time of detection of zero-sequence components is applied as the time of wave arrival. In this thesis, a laboratory setup has been developed to show the problems of existing relays in TW based fault detection. Thereafter, technical detail of zero-sequence component detection based on the Current Transformers (CT) and Capacitor Voltage Transformers (CVT) data with the algorithm of estimation to fault has been introduced.*

## 2.3 Transmission Line Protection Functions

Power system protection deals with identification and isolation of faults occurring in the electrical network [43]. The main aim of power system protection is to safeguard power system equipment such as transformers, generators, transmission lines, feeder etc, from faults. Moreover, since overhead transmission lines are exposed to the environment, there exists a greater possibility of occurrence of faults in overhead transmission lines than any other power system equipment [44]. Due to increased electrical energy consumption and growing impetus towards distributed generation

[45], fluctuating power flows and varying levels of short circuit currents occur in the electrical grid [46]. Differential and distance relays with remote backup protection have been implemented as the main transmission line protection functions for many years [47]. Following subsections review the literature of these protection functions in transmission lines.

### 2.3.1 Distance Protection

Distance protection scheme is one of the most sought over technique by relay engineers for reliable protection of transmission line because of its widespread usage in power systems and availability of advanced distance protection computational algorithms. Distance relay calculates real time impedance and compares it with the basic impedance of transmission lineand if it is found to be less; the fault is determined to occur in the reach of the relay. Then, a trip command is issued to the local circuit breaker to clear the fault [48, 49]. The main problem of this technique is that the technique gets affected by external disturbances. The reports of NERC (North American Electric Reliability Corporation) show that the mal-operation in distance relays is one of the important reasons of tripping in power system [50]. With the development of modern power systems, the distance protection is becoming more and more complicated. The performance of distance protection for connecting a line to HVDC [51, 52], FACTS devices [53-60] and wind farms [61-64] have been the main avenue of researches over the last few years. Based on the literature review, some problems in transmission line protection functions can be summarized as bellow:

- Over Reach and Under Reach Distance Relay

In distance relay over reach, the effect of another source into fault point in multi-terminals which distance relays measure higher impedance than the actual which can cause delay to trip. In distance relay under reach, external faults are fed from many terminals. In this situation, the distance relay faces with lower impedance than the real impedance to fault and it may false trip would be happened [65].

- Load Encroachment and Unbalanced Load

When the load is changing, it may meet the relay trip zone and operate as a fault. The unbalanced nature of loads and asymmetry of lines can affect the performance of distance protection. Therefore, there is low precision in protection zone limits of ground distance relays [66]. A relay uses a load-encroachment feature that prevents operation of the phase distance elements during heavy load. This feature permits the load to enter a pre-defined area of the phase distance characteristics without causing unwanted tripping. Figure 2-6 shows the settings of load-encroachment function in the SEL 421 relay.

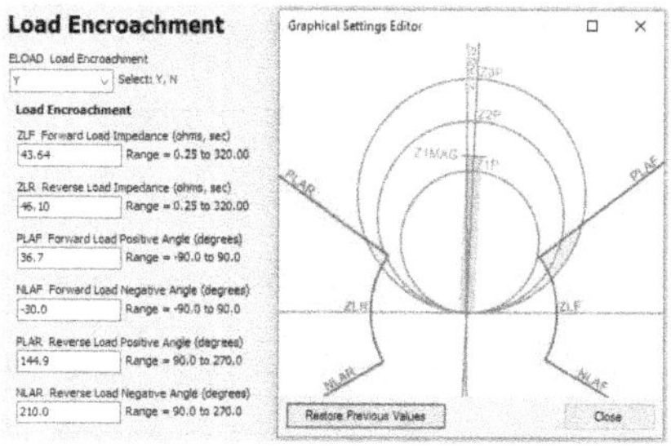


Figure 2-6- Load encroachment relay setting

Figure 2-7 shows the load-encroachment logic. The load encroachment logic prevents load from causing phase protection to operate. Two independent positive sequence impedance characteristics monitor the positive sequence load impedance for both export and import load. The positive sequence voltage-polarized directional element is blocked when the load encroachment logic is enabled and load is detected. The phase distance elements cannot operate during balanced system conditions unless the logic asserts the directional element [67].

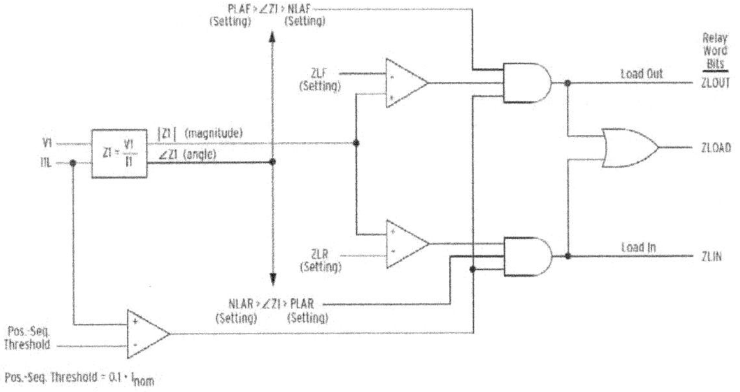


Figure 2-7- Load encroachment function logic [67].

- Power Swing

In normal conditions, power system operates near nominal frequency. In other words, there is balance between generated power and consumed load. In some condition, there is a sudden change in electrical system due to faults, switching and connection or disconnection of the bigger loads. In this case, the measured impedance by the distance relay may meets its protected zone and false trip may happen. Power swing block which is shown in Figure 2-8, is a function in distance relay to prevent unwanted operation of relay during swings. However, faults that occur during a power swing have to be detected and isolated very fast. In these circumstances, the power swing block should be unblocked. It means the distance relay should operate to clear the fault [68-72].

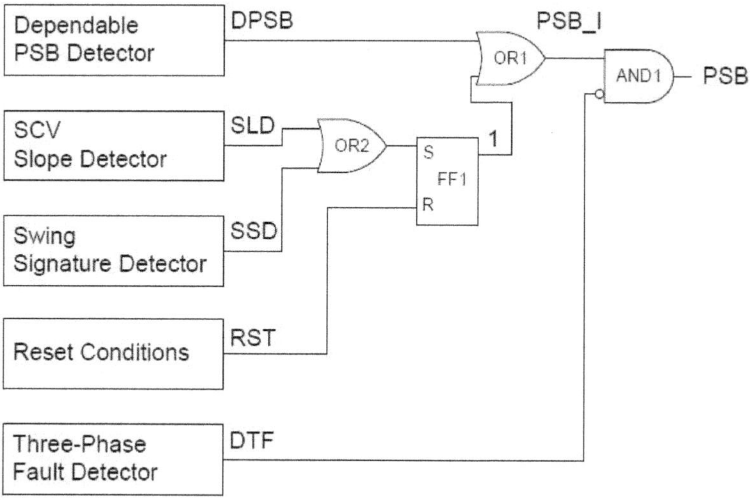


Figure 2-8- Power swing logic function [67].

Based in a lab setup, out-of-step blocking is the function in the SEL-421 relay which blocks phase distance elements and ground distance elements during power swings when the measured positive sequence impedance enters the operating characteristics of the phase distance elements. In practice, it is not necessary to block all protection zones. Therefore, in the relay setting, instantaneous part of first and second zones should be activated to prevent overreaching protection zones that provide time delayed tripping.

- Mutual Impedance Between Parallel Lines

Parallel transmission line or double-circuit transmission lines have the advantages of high power transfer capacity and also economic benefits over single-circuit lines. However, the fault on them is unavoidable. The fault between circuits and the mutual coupling between lines are the main difficulties to protect that kind of lines. The mutual coupling between the lines can be affected on the performance of a distance relay, especially for cross-country faults [73-76].

- High Resistance Faults

The grounded fault current depends on the resistance of arc, tower and the ground system. In high impedance fault (HIF), distance relay send the trip signal with delay [77-79]. Several methods have been investigated to detect a high impedance fault in transmission lines [80-82].

- Mal-operation of Zone 3

One of the main reasons of large disturbances in power systems is mal-operation of the zone 3 in distance relays especially in stressed conditions, such as power swing, voltage stressed condition and load encroachment. The mal-operation of distance relays decreases the reliability of system [83-85].

### 2.3.2 Accelerated Distance Protection

Conventional distance protection does not provide instantaneous tripping for all types of faults on transmission line [86, 87]. The remote faults are normally cleared by the conventional distance protection scheme after the Zone-II time delay. The fault conditions during this time delay may lead to system instability or cascading blackout.

Also from research, it is evident that for every one millisecond saved in fault clearing time, the stability limit of the transmission system goes up by 15 MW [88]. Hence the time delay for the Zone-II faults has to be decreased. One of the solutions is to provide communication assisted accelerated distance protection scheme for Zone-II faults in the transmission line which can speed up the overall fault clearing time [89].

In order to have this communication assisted accelerated distance protection scheme through standardized and interoperable means, IEC 61850 [90] based communication has been widely adopted [91, 92]. IEC 61850 has emerged as a de-facto standard for substation automation in utilities world over [93]. In recent years, IEC 61850 based substation automation research activities have been well reported in the literature and still substantial research is underway [94-96]. IEC 61850 standard was initially proposed for substation automation which was later extended to utility automation [97]. With the publication of new parts such as IEC 61850-90-1 [98], it has been extended for substation to substation (inter-substation) communication. Hence IEC 61850 based communication has been used in different protection schemes between different substations [99, 100]. The advantage of adopting IEC 61850 for accelerated distance protection scheme is to achieve fast tripping for Zone-II relay based on transportation of the high speed GOOSE messages from the distant relay.

### 2.3.3 Differential Protection

Differential protection is a reliable protection method and it is the best protection method in power system [101]. The charging current of capacitance is the main problem of differential protection in the long lines, whereas, it can be neglected in short lines. The authors [102] have proposed a combination of impedance and differential scheme to remove the impact of the charging current. Researchers have proposed to apply the skewnesses of the differential current and restrain current which does not need to compensate the captive current [103]. Transmission line differential protection function based on Phasor Measurement Units (PMUs) has also been proposed in [104, 105]. Another work is based on travelling wave has presented in [52, 106]. Many algorithms have been developed to deal with the impacts of the charging current compensation [107-110]. Most of them apply a higher operating curve which decreases the sensitivity of line differential protection.

- *The performance of the distance relay is affected by external disturbances in line such as power swing, fault resistance, FACTS devices, etc. Differential protection is a reliable protection method to secure the protection zone. However, its performance is not satisfactory in long TLs. In this thesis, a new scheme based on distributed current in a line is proposed. A new protection function is developed based on these measured values and the performance of the new scheme is compared with existing differential protection.*

## 2.4 IEC 61850 Standard

IEC 61850 is the international standard for communication in power network. The main goal of the standard is to provide interoperability. In IEC 61850 all functions have been identified and split into sub functions, called logical node (LN). The may be grouped into Logical Devices (LDs). LDs are implemented in Intelligent Electronic Devices (IEDs) [111, 112]. The IEC 61850 standard has many benefits such as replacing the traditional copper wiring with GOOSE messages that are sent via Ethernet over fibre optic cables [112]. Several functions in substation automation based on IEC 61850 had been presented in [113-117]. To describe a unified function model of IEDs, an approach of combining of IEC 61850 with IEC 61499 have been proposed in [118, 119]. The End-to-End (ETE) time delay has been evaluated using OPNET simulation package in [93, 120], and the utilization of high-speed peer-to-peer communications, multicast sample values [121] and time synchronization with IEEE 1588 [122, 123] allows the introduction of distributed and wide area in smart grids.

- *By literature review, it was found that there is a gap between practical implementation of IEC 61850 and research. This thesis presents the practical consideration of IEC 61850 in SAS at the substation automation laboratory at JMI.*

- *To implement the distance protection between two substations based on IEC 61850, the GOOSE messages have to be transported from one substation to another distantly located substation. But the GOOSE messages are typically mapped on to data link layers, hence cannot be transported over a Wide Area Network (WAN). IEC 61850-90-1 recommends tunnelling of GOOSE messages over a WAN for inter-substation communication-based protection schemes. Many authors have explored inter-substation communication, but performance evaluation of tunnelled GOOSE messages over the WAN under different scenarios has not been reported. This work explores these methods and presents an evaluation of different methods for different protocols using software in loop simulation results. Also, in order to tunnel GOOSE messages for inter substation communication; there is a challenge of configuring the different IEDs located in different substations. Hence, the details on process of exchange of SCL files for configuring IED in different substations are presented in chapter 5.*

## 2.5 Reliability Investigation of SAS

It is using IEC 61850 standard which is the international standard for communications within substations. Furthermore, the method to transmit the current and voltage data at the process level using a network has been standardized within IEC 61850 [124]. However, it is necessary to select each component and the configuration of SAS from the reliability point of view. To improve the reliability, it is very common to use the redundant equipment in SAS. Moreover, most of the time an important substation needs the high level of redundancy in the SAS network [125]. As the reliability of SAS is interested by utilities and manufacturers, various research works have been considered on the reliability of SAS [126, 127]. The assessment of reliability and availability in different topology of substation automation system has been calculated in [128-130]. The main objective of IEC 61850 is to facilitate the interoperability between different manufacturers. It enables the connection different types of equipment from several vendors through an Ethernet LAN [131].

- *IEC 61850 does not standardise the redundant configuration and each manufacturer uses different topology in SAS to increase the reliability and at the same time reduces the cost. The reliability and availability of several topologies for a feeder protection are assessed in chapter 6.*

## 2.6 Conclusion

In this chapter, a comprehensive review of the development in the area of transmission line operation and protection functions using communication technology has been presented. From the literature, it is observed that communication technology is integrating to electrical automation system rapidly. However, there is a gap between research and industry in this area. Therefore, more research with more practical concentration is needed in the area.

In the next chapter, an overview of fault location in the transmission line is discussed along with the development of impedance-based fault location using communication technology and integration of travelling wave based fault location into the impedance-based relays.

# Chapter 3 New Schema in Impedance- and Travelling Wave-based Fault Location in Transmission Line

## 3.1 Introduction

A transmission line fault analysis tool should detect and estimate the location of the faults in the transmission line by interpreting of the transient fault data. This chapter reviews research work done on the existing transmission line fault location algorithms. New impedance-based fault location methods based on distributed current and voltage in line have been introduced in this chapter.

---

The content of this chapter is based on the following papers:

1- Saeed Roostaee, Mini S Thomas, and Shabana Mehfuz "Laboratory Investigation of Fault Location in Transmission Lines "Journal of Electrical Engineering, Electronics, Control and Computer Science 2.2 (2016).
2- Saeed Roostaee, Mini S Thomas, and Shabana Mehfuz, "Experimental studies on impedance based fault location for long transmission lines ", Journal of Protection and Control of Modern Power Systems, no. 1 (2017): P.16 www.springeropen.com/journals, 2017
3- Saeed Roostaee, Shabana Mehfuz, and Mini S Thomas, "Traveling-Waves-Based Ground Fault Location Using Zero-Sequence Detection and Wavelet Transform" Journal of Electrical Engineering, Electronics, Control and Computer Science 1.1 (2017): 7-12.
4- Saeed Roostaee, Shabana Mehfuz, and Mini S Thomas," Impedance Based Fault Location in the Power Transmission Line based on IEC 61850 Process Bus", IEEE Nanotechnology for Instrumentation and Measurement Workshop, IEEE NANOFIM 2017

The feasibility of the new scheme is simulated in PSCAD. The performance of the proposed schema is evaluated with experimental results.

To integrate TW-based fault location with existing protection relay, a new travelling wave-based method is proposed to estimate the location of ground faults.

## 3.2 Impedance-based Fault Location with Distributed Measured Values in a Transmission Line

Impedance-based fault location algorithms are used to estimate the location of faults on a transmission line. These methods are commonly used by utilities because of simplicity. They require current and voltage signals along with sequence impedance (positive, negative and zero) to estimate the distance to fault location. Depending on the availability of the input signals, these algorithms can be categorized into one-end and two-end methods. Following subsections provide an explanation of both the methods.

### 3.2.1 One-end Impedance-based Fault Location Algorithm

In this technique, the ground phase voltage and current are required. Figure 3-1 shows the source at one end where $I_F=I_A$ and the estimated distance can be calculated from Eq. 3.2 [5].

$$Z_F = \frac{U_A}{I_A} = dZ_L + R_F$$ Eq. 3.1

$$d = \frac{\operatorname{Im}(Z_F)}{\operatorname{Im}(Z_L)}$$ Eq. 3.2

Where $Z_F$ is the measured impedance from substation A, $Z_L$ is the impedance of the line, $R_F$ is the resistance of fault.

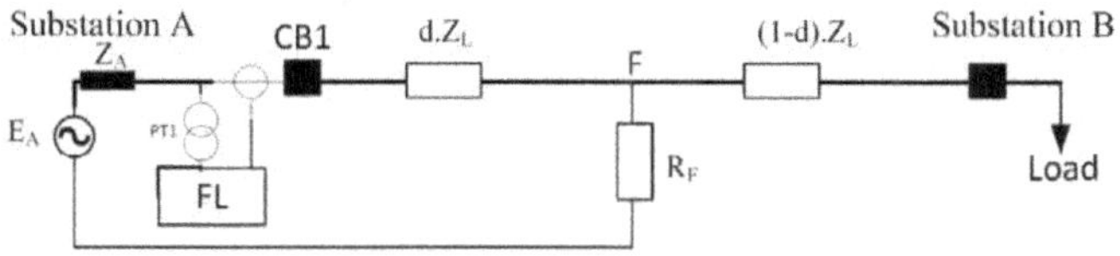


Figure 3-1- Impedance-based fault location with the source at one end

### 3.2.2 Two-end Fault Location Algorithm

Figure 3-2 shows impedance-based fault location based on the source at two ends, where $I_F = I_A + I_B$. In these types of lines, estimated distance can be obtained from Eq. 3.4.

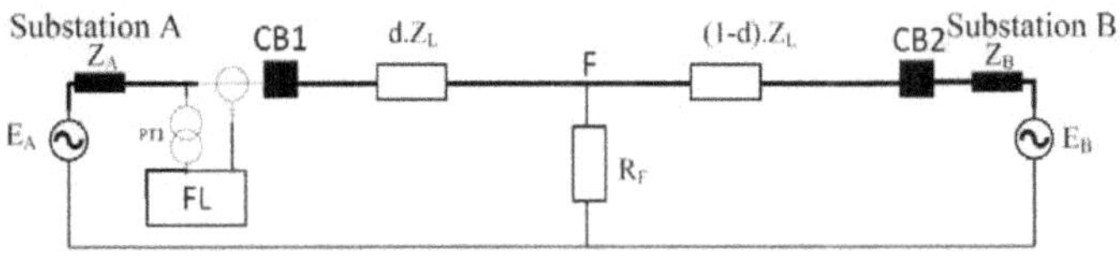


Figure 3-2- Impedance-based fault location with the source at two ends

$$U_A - dZ_L I_A - R_F I_F = 0 \quad \text{Eq. 3.3}$$

$$d = \frac{\mathrm{Im}(U_A \Delta I_A^*)}{\mathrm{Im}(Z_L I_A \Delta I_A^*)} \quad \text{Eq. 3.4}$$

In two ends impedance-based fault location, the voltage of fault from each end is equal.

$$U_{Fi}^A = U_{Fi}^B \quad \text{Eq. 3.5}$$

The voltage of fault from each end can be obtained from Eq. 3.6 [5].

$$\begin{aligned} U_{Fi}^A &= \cosh(\gamma_i dl) U_{Ai} - \sinh(\gamma_i dl) Z_{Ci} I_{Ai} \\ U_{Fi}^B &= \cosh(\gamma_i (1-d) l) U_{Bi} - \sinh(\gamma_i (1-d) l) Z_{Ci} I_{Bi} \end{aligned} \quad \text{Eq. 3.6}$$

Where $U_{Ai}$ and $U_{Bi}$ are the i-th symmetrical component of voltage in A and B respectively, $I_{Ai}$ and $I_{Bi}$ are the i-th symmetrical component of current in A and B respectively. These values can be obtained by instrument transformers at both the ends of a line. Following trigonometric identities can simplify the Eq. 3.6.

$$\begin{aligned} &\cosh(\gamma_i (1-d) l) = \\ &\cosh(\gamma_i l)\cosh(\gamma_i dl) - \sinh(\gamma_i l)\sinh(\gamma_i dl) \\ \\ &\sinh(\gamma_i (1-d) l) = \\ &\sinh(\gamma_i l)\cosh(\gamma_i dl) - \cosh(\gamma_i l)\sinh(\gamma_i dl) \end{aligned} \quad \text{Eq. 3.7}$$

To estimate the location of the fault based on two ends values, following formula can be derivate from Eq. 3.6 and Eq. 3.7 [5].

$$d = \frac{1}{\gamma_i l}\tanh^{-1}\left(\frac{\cosh(\gamma_i l)U_{Bi} - Z_{Ci}\sinh(\gamma_i l)I_{Bi} - U_{Ai}}{\sinh(\gamma_i l)U_{Bi} - Z_{Ci}\cosh(\gamma_i l)I_{Bi} - Z_{Ci}I_{Ai}}\right) \quad \text{Eq. 3.8}$$

Where $\gamma_i$ is propagation constant of the line for i-th symmetrical component, $Z_{Ci}$ is the characteristic impedance of the line for i-th symmetrical component.

### 3.2.3 Proposed Scheme

Generally, a fault location algorithm uses the recorded transient data from the protection relays at both the ends of a line. In this study, we propose to use transient data of the protection relays in substations as well as transient data of the DFRs which are installed in a line. As it is shown in Figure 3-3, the DFRs are installed in a transmission line and they can send transient fault data to both substations through a communication channel.

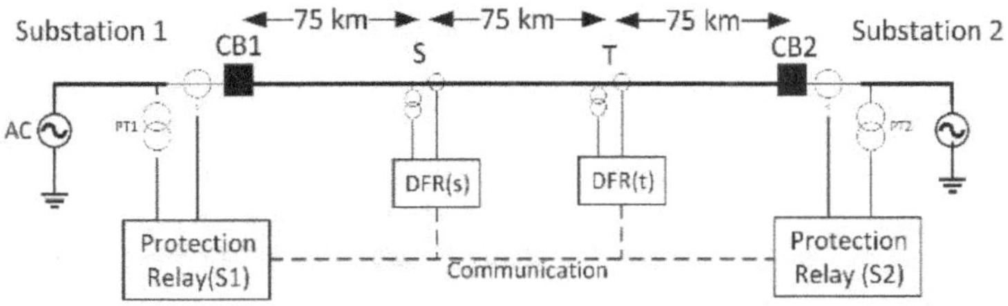


Figure 3-3- Proposed scheme

A laboratory test bench is developed in order to be able to implement and test the proposed fault location method. Figure 3-4 shows the build test bench at Substation Automaton laboratory at Jamia Millia Islamia.

Figure 3-4- The lab test bench at Jamia Millia Islamia

Figure 3-5 illustrates the schematic diagram of the developed lab where a PC running PSCAD is connected to the CMC 256-6 plus, and controls the inputs/outputs of the CMC 256-6 plus. The voltage and current outputs of the CMC 256-6 is wired to the SEL 421. Hence, we can simulate transmission line faults in PSCAD; after that, the transient data for the fault is loaded to the CMC 256-6. The CMC can playback the transient values and generate the real three phase's voltage and current. These three phase signals are connected to the SEL-421 relay throughout wiring.

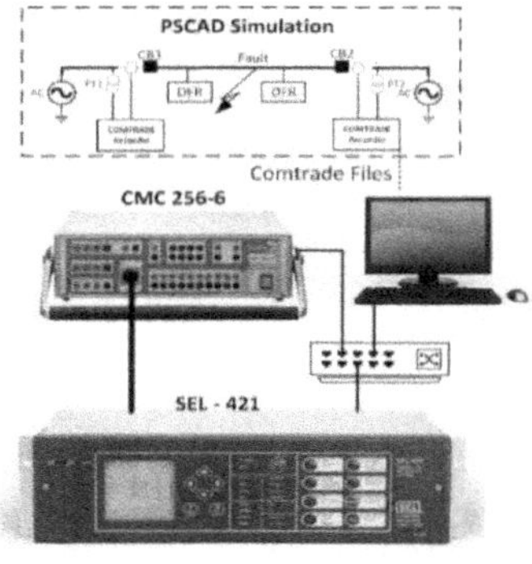


Figure 3-5- Schematic diagram of the lab

Based on this setup, various types of faults in different locations of a transmission line can be simulated and the associated fault data can be generated via PSCAD (Figure 3-6). Transient data can describe the voltage and current signals both before and after the fault. There is a component in PSCAD which can record the transient data in the COMTRADE format. This format is an IEEE standard format for recording transient data and it can be loaded with different fault analysis tools [132].

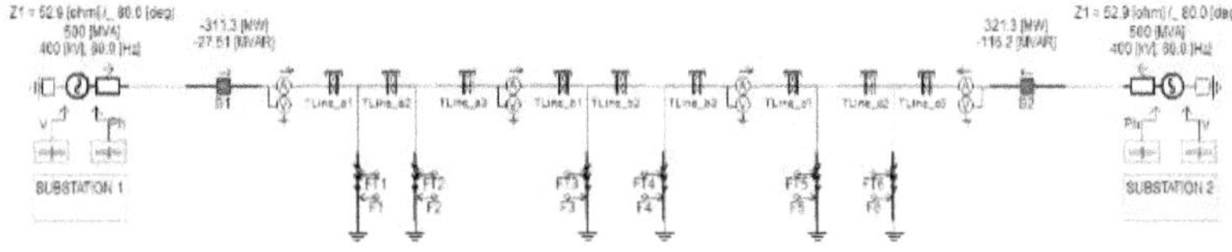


Figure 3-6- PSCAD Simulation

The CMC 256 test set can generate arbitrary three-phase signals. Many applications such as testing synchronizer function, protection functions can be implemented with this test set. One of the most interesting features of this test set is that it can load the transient data and playback of them to inject the related three phase voltage and current to an object. The Transview is a tool which can load COMTRADE file and generate corresponding three phase signals (Figure 3-7). The signals can be utilized by a protection relay IED to test a fault location algorithm.

Following subsections discuss a case study and evaluate the proposed scheme for the different type of faults in different locations.

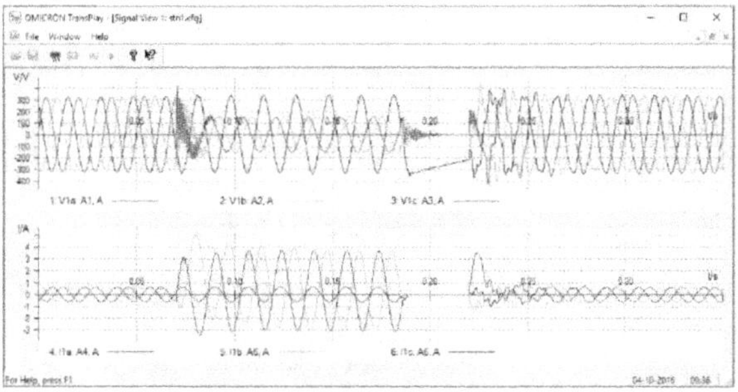

Figure 3-7- Trans-View play

### 3.2.4 Simulation of Power System Model in PSCAD/EMTDC

Modelling of a transmission line is very important in fault location algorithms. Impedance-based fault location technique estimates the location of a fault by comparison of impedance with the line impedance. Therefore, the accuracy of this technique highly depends on the accuracy of line parameter model. To test the proposed scheme, we have simulated a 400 kV transmission line between substation 1 and substation 2 which is a homogeneous and un-transposed transmission line. In this study, standard tower, conductor, and ground wire have been selected (Figure 3-8). Figure 3-9 shows the main feature of the HS10 tower which is simulated in PSCAD. To transfer the power, each phase contains four conductors in symmetrical bundle mode with 45.72-centimetre spaces between two connected conductors. To protect the line against lightning as well as communication purpose, OPGW-24B1/70 is selected as the ground wire. The main features of conductors and ground wires are summarized in Table 3-1.

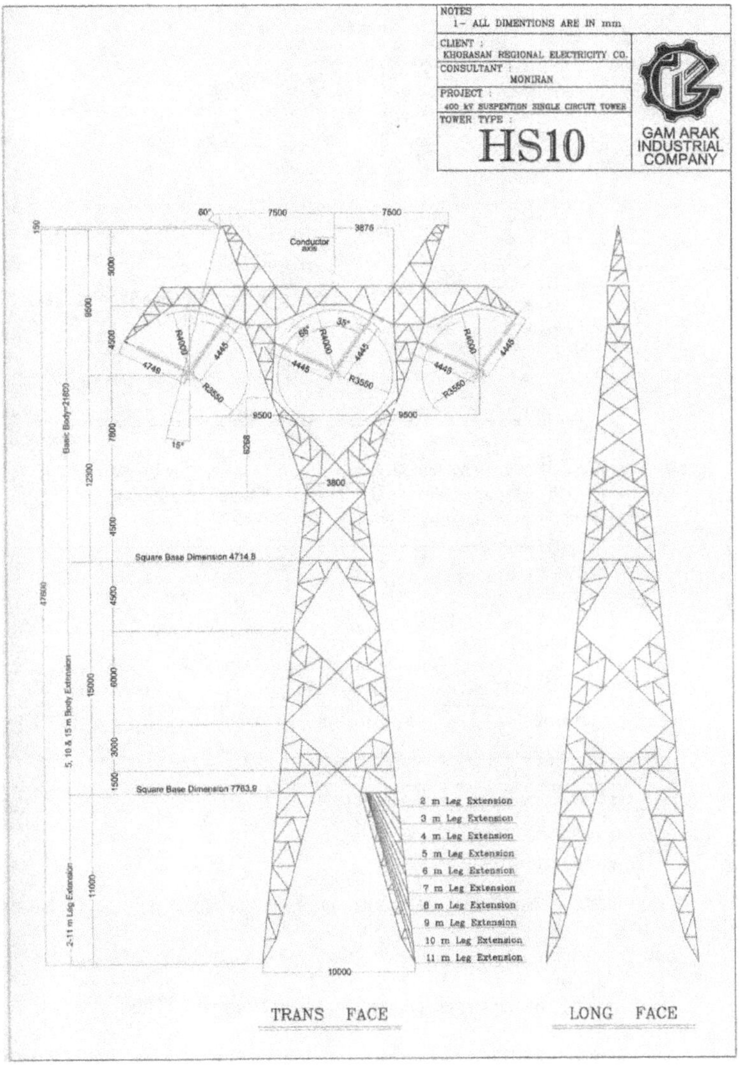


Figure 3-8- Selected tower for simulation

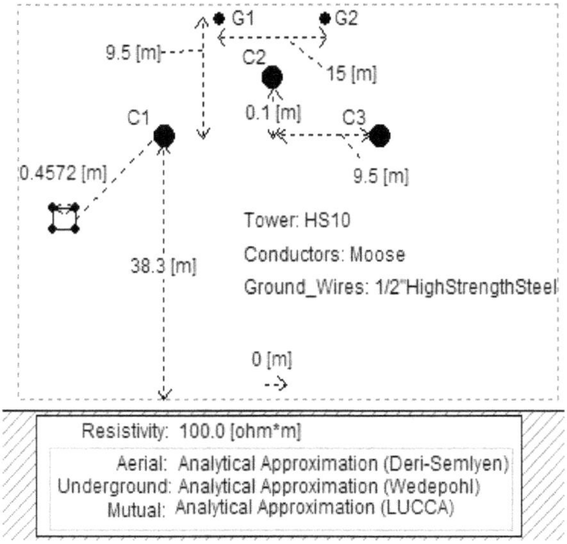


Figure 3-9- 400kV transmission line tower

Table 3-1- Conductor and ground wire

| Name | Application | Diameter | $R_{DC}$ (20°C) |
|---|---|---|---|
| | | mm | Ω/km |
| ACSR- Moose | Conductor | 31.77 | 0.0547 |
| OPGW-24B1/70 | Ground cable | 12 | 0.79 |

Distributed transmission line modelling is used to model the line. In this model, the line parameters can be calculated based on the Carson's' equations [133] and the characteristics of the tower, conductor, and ground wire. The resistance, inductance, capacitance, as well as positive and zero sequences of resistance and inductance were calculated in our study as below:

$$R = \begin{pmatrix} 0.038906 & 0.02464 & 0.024644 \\ 0.02464 & 0.038614 & 0.02464 \\ 0.024644 & 0.02464 & 0.038906 \end{pmatrix} \Omega / km$$

$$L = \begin{pmatrix} 0.0013674 & 0.00053991 & 0.00041175 \\ 0.00053991 & 0.0013513 & 0.00053991 \\ 0.00041175 & 0.00053991 & 0.0013674 \end{pmatrix} \text{H}/km$$

$$C = \begin{pmatrix} 1.0871e-8 & -2.734e-9 & -1.0124e-9 \\ -2.734e-9 & 1.1534e-8 & -2.734e-9 \\ -1.0124e-9 & -2.734e-9 & 1.0871e-8 \end{pmatrix} F / km$$

$$\begin{bmatrix} R1 & R0 \end{bmatrix} = \begin{bmatrix} 0.014167 & 0.088091 \end{bmatrix} \Omega / km$$

$$\begin{bmatrix} L1 & L0 \end{bmatrix} = \begin{bmatrix} 0.00086486 & 0.0023564 \end{bmatrix} \text{H}/km$$

To evaluate the proposed scheme, several faults in different location are simulated which are discussed in the following subsections.

### 3.2.5 Faults Away from the Relay Location

This subsection discusses the faults between the S and T points in the proposed scheme (Figure 3-3). To test the scheme, a transient fault with following features is simulated in PSCAD.

- Fault Location: 125 km
- Fault type: AB-g (A-B-G)
- Fault Start: 0.22 s
- Fault duration: 0.15 s
- Fault Resistance: 0.001 Ohm

The transient data for the above fault is analysed in OMICRON-Trans View software. As it is shown in Figure 3-10, a screen capture of the software, the estimation of 130.9 km far from substation 1 is the estimation of the location of the fault.

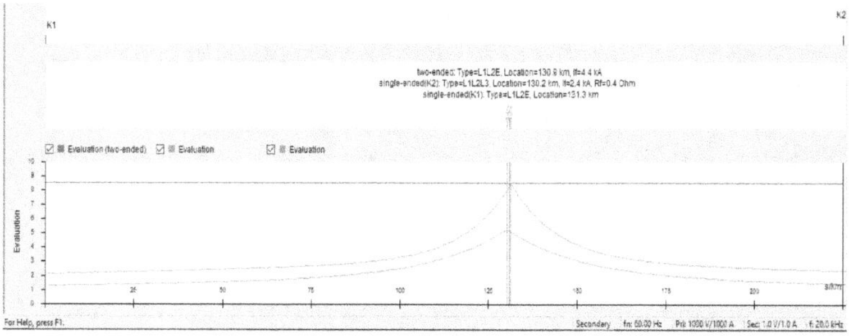


Figure 3-10- Both ends fault location

The SEL-421 is a transmission line relay and it has five zones of phase and ground. This relay has online fault location and after a fault occurs in a transmission line, the estimation of fault location is shown in relay' LCD as well as, the transient fault data along with estimated fault location are recorded as an event in the relay. The setting of the relay is important in fault location function. Based on the data of the transmission line, following settings have been set in the relay:

- Frequency: 60 Hz
- Line Length: 225
- Positive Impedance Magnitude: 73.4
- Positive Impedance Angle: 87.51
- Zero Impedance: 200.76
- Zero Impedance Angle: 84.3

After injection of transient data of the fault to the relay by CMC 256, the relay issues a trip signal and record the transient data of the fault (Figure 3-11). The relay detects an AB-g fault in zone one. Thereafter, online relay fault location estimates a fault on 135.38 km far from substation 1 (Figure 3-12).

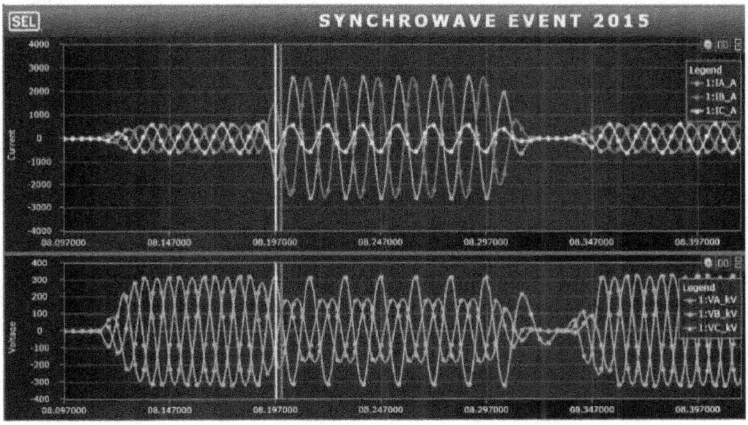


Figure 3-11- Relay transient data record

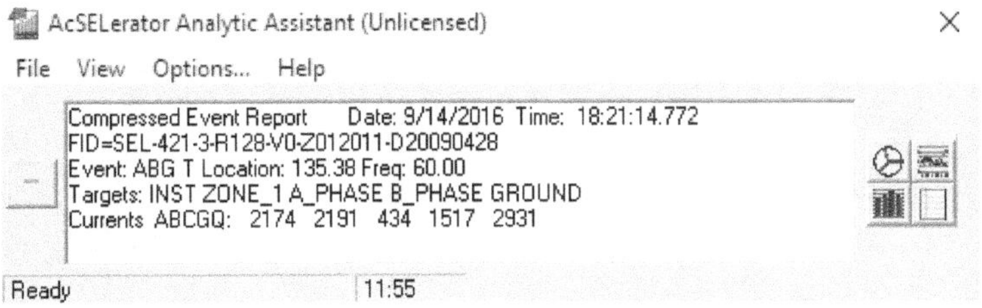


Figure 3-12- Relay event view

In the proposed scheme, the recorded event of DFR1 and DFR2 are utilized to estimate the fault location. Based on the proposed scheme which is described in 3.2.3, a transmission line is divided into three sections. Section one is between substation one and DFR1; section two is between DFR1 and DFR2; and the last section in between DFR2 and substation two. For the AB-g fault in the middle of the line, only section two has detected the fault. The fault location estimation is 49.8 km far from DFR1 (Figure 3-13). In the other words, the estimated location is 124.8 km far from the substation one.

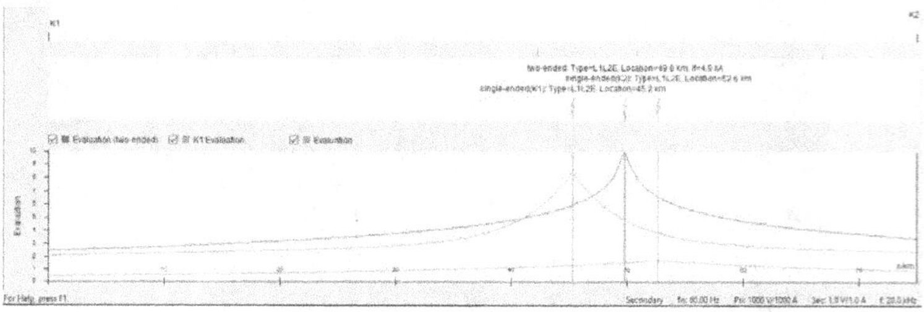

Figure 3-13- Fault location based on the proposed scheme

To compare the accuracy of fault location based on different methods, the percentage error is determined as:

$$\%Error = \frac{L_{Estimation} - L_{Actual}}{L_{Line}} * 100 \qquad \text{Eq. 3.9}$$

Where $L_{Estimation}$ is the estimated location by fault location algorithm, $L_{Actual}$ is the actual location of the fault and $L_{Line}$ is the line length.

The accuracy of different techniques for the fault which is simulated in this subsection can be summarized in Table 3-2.

Table 3-2- Comparison of accuracy with different fault location method

| Fault Type | $R_{Fault}$ | Location | Fault Location Error (%) | | |
|---|---|---|---|---|---|
| | Ω | km | With two ends method | Relay Fault Location Function | With the proposed scheme |
| AB-G | 0 | 125 | 2.6222 | 4.6133 | -0.089 |

Based on this experiment, it was observed that the proposed technique is 29 and 51 times more accurate than two ends and relay function based fault location respectively. Similar observations can be archived for other faults, which are summarized in Table 3-3.

Table 3-3- Comparison of accuracy of different faults between point "s" and point "t"

| Fault Type | $R_{Fault}$ Ω | Location km | Fault Location Error (%) | | |
|---|---|---|---|---|---|
| | | | With two ends method | Relay Fault Location Function | With the proposed scheme |
| AB | 0 | 100 | 3.4222 | 5.1733 | 0.2222 |
| AB | 0 | 125 | 2.7111 | 4.8311 | 0.0444 |
| AB | 50 | 100 | 3.3333 | 4.9333 | 0.6667 |
| AB | 50 | 125 | 2.6222 | 4.6133 | -0.089 |
| ABG | 0 | 100 | 3.3333 | 4.9333 | 0.6667 |
| ABG | 0 | 125 | 2.6222 | 4.6133 | -0.089 |
| ABG | 50 | 100 | 3.2444 | 139.2133 | 0 |
| ABG | 50 | 125 | 2.5333 | 143.2489 | -0.356 |
| BG | 0 | 100 | 3.1111 | 6.6488 | 0.6667 |
| BG | 0 | 125 | 2.2667 | 7.1155 | -0.622 |
| BG | 50 | 100 | 2.8444 | 8.8933 | 0.3556 |
| BG | 50 | 125 | 2.1333 | 10.6889 | 0.3556 |

### 3.2.6 Faults near the Relay Location

In case of failure of the measurement system in a substation, both ends methods cannot be applied. Thereafter, the measured values from the remote end can estimate the fault location with one-end algorithms. But, in the proposed scheme, transient data of DFRs which is installed in the line can be applied.

To evaluate the proposed scheme, several faults with different locations, types, and resistances are simulated. Testing of various fault location techniques are carried out by utilizing the hardware implementation in a laboratory environment. The result obtained from the preliminary analysis of the accuracy of different techniques are summarised in Table 3-4. Based on this experiment, it is observed that the proposed technique is more accurate than one-end fault location where the proposed technique applies measured value by DFRs in line, but the existing method only applies the measured values in substation.

Table 3-4- Comparison of accuracy of different faults between point "t" and substation b

| Fault Type | $R_{Fault}$ Ω | Location km | Fault Location Error (%) | |
|---|---|---|---|---|
| | | | One-end | Proposed shceme |
| AB | 0 | 175 | 4.4888 | 0.3111 |
| AB | 0 | 200 | 4.5066 | -0.089 |
| AB | 50 | 175 | 9.9644 | 0.8 |
| AB | 50 | 200 | 12.44 | 0.2667 |
| ABG | 0 | 175 | 2.24 | 0.5778 |
| ABG | 0 | 200 | 4.4044 | 0.2222 |
| ABG | 50 | 175 | 191.87 | 0.8 |
| ABG | 50 | 200 | 228.4 | -0.089 |
| BG | 0 | 175 | 8.2133 | 0.7111 |
| BG | 0 | 200 | 8.7911 | 0.2222 |
| BG | 50 | 175 | 17.1289 | 0.4 |
| BG | 50 | 200 | 23.3778 | 0.4 |

## 3.3 Travelling-Waves-Based Ground Fault Location Using Zero-Sequence Detection

Statistics show 70-80 percent of faults in transmission lines are single line to ground where they can pose a serious threat to the safety of power system. Hence, fault location for these types of fault is very important. This thesis proposes to apply the time of zero sequence detection in TW-based fault location method in ground faults. Subsequent sections discuss the basic and the details of the proposed technique in TW-based fault location.

### 3.3.1 Description

When a fault occurs on transmission lines, impulse signals or travelling waves are generated. The waves can be monitored in the both end of a transmission line. Ground fault is one of the most common reasons of generation of TW where the wave of voltage and current can be described by Eq. 3.10 and Eq. 3.11.

$$e(x,t) = e_f(x - vt) + e_r(x + vt) \quad \text{Eq. 3.10}$$

$$i(x,t) = \frac{1}{Z} e_f(x - vt) - \frac{1}{Z} e_r(x + vt) \quad \text{Eq. 3.11}$$

Where $e$ is voltage wave, $i$ is current wave, $e_f$ and $e_r$ are equation of forward wave and backward wave respectively, $Z$ is the impedance of line, $v$ is wave velocity, $t$ is time, and $x$ is the distance from the measuring point.

Based on the velocity and travelled time, the estimation of fault location can be obtained from Eq. 3.12.

$$x = \frac{v\mathrm{T}}{2} \qquad \text{Eq. 3.12}$$

Where T is signal travel time, and x is estimated fault location.

To improve the performance of this technique, many scholars have proposed a number of methods and techniques. The authors [134] have used wavelet transform. Transient fault data are usually non stationary signals. In the other words, their frequency contents change with time. Wavelet transform is a powerful method to analysis such signals. Based on the wavelet theory, signals can be decomposed into a series of wavelets from equation Eq. 3.13.

$$\Psi_{S,\tau}(t) = \frac{1}{\sqrt{S}} \Psi\left(\frac{t-\tau}{s}\right) \qquad \text{Eq. 3.13}$$

Where $\psi(t)$ is mother wavelet, $S$ and $\tau$ are scale and translation parameter, and the term $1/\sqrt{S}$ is utilized to normalize wavelet energy of different scales. Wavelet transform can be obtained by implementing discrete wavelet transform which is given by;

$$\Psi_{j,k}(t) = 2^{-j/2}\psi(2^{-j}\,\mathrm{t}-\mathrm{k})$$ Eq. 3.14

Where j is scale parameter, k is translation parameter, j is a parameter which can change the amount of signal compression, and k displaces the wavelet in time domain.

Detection of surge is one of the most important factors in TW-based fault location methods. One of a desirable method to detect a ground fault is to detect the zero-sequence components which is proposed in this thesis to detect the travelling wave surge. The detection of zero sequence strategy is discussed in the subsequent sections.

### 3.3.2 Zero-Sequence Detection in Ground Fault

In normal condition, the sum of the three phase currents/voltages is zero resulting in the zero output for zero sequence detection. When a ground fault happens in the system, the sum of the three phase currents/voltages is not equal to zero. Following subsections explain the detection of zero-sequence by CT and CVT.

#### 3.3.2.1 $I_0$ Detection by CT

Figure 3-14 shows a zero sequence CT. And Figure 3-15 shows the frequency response for different types of CTs. As the CTs have wider band-width than CVTs, it is perfectible to apply CTs in the proposed fault location algorithm. The zero sequence detection based on CT is as follow:

$$I_0 = I_A + I_B + I_C \qquad \text{Eq. 3.15}$$

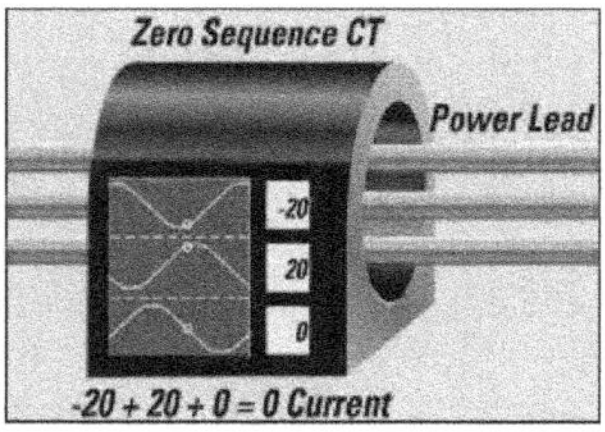


Figure 3-14 Zero sequence CT

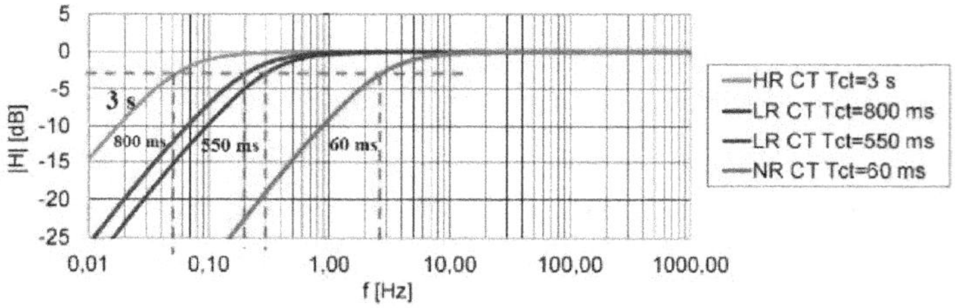


Figure 3-15 Frequency response for different CT types

#### 3.3.2.2 $I_0$ Detection by CVT

Capacitor Voltage Transformers is an instrument transformer which is used in substations for voltage measurement. It converts high voltage to low voltage for substation automation and protection functions. Capacitor Voltage Divider (CVD) and electromagnetic transformer are the main part of a CVT. CVD convert the voltage level and electromagnetic transformer changes the voltage level to the standard values. Figure 3-16 shows the model of a CVT. There are two outputs in this model. One output is used by Power Line Carrier (PLC) for communication purpose and the other one is used by automation and protection systems.

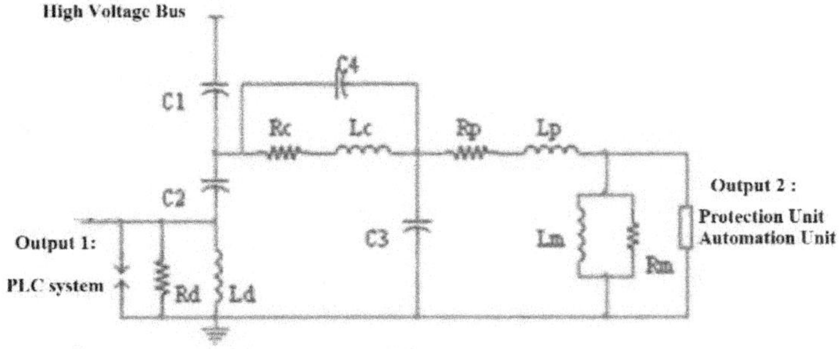


Figure 3-16- CVT model

The frequency response of the output 1 and output 2 are shown Figure 3-17 and Figure 3-18 respectively. Terminal 2 can pass high frequency with a higher gain compare to terminal 1. Therefore, TW would pass through terminal 2. To detect zero sequence component based CVT, a current sensor is installed on the terminal 2 of a CVT, which can detect zero sequence component as follow:

$$I_0 = C\frac{dV_E}{dt} \quad \text{Eq. 3.16}$$

Where C is the capacitance of CVT, and $V_E$ is the phase voltage.

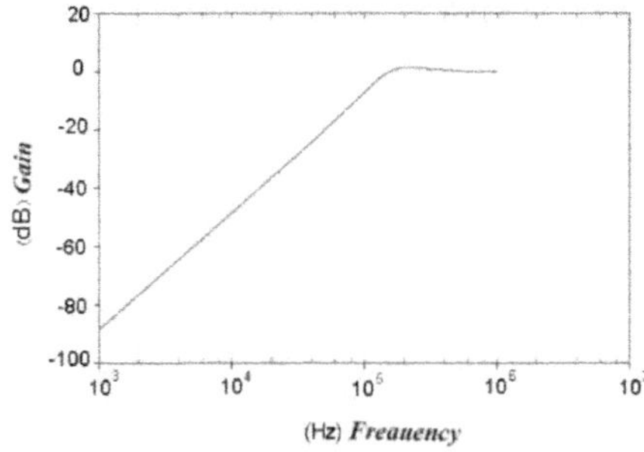


Figure 3-17- Frequency response of output 1

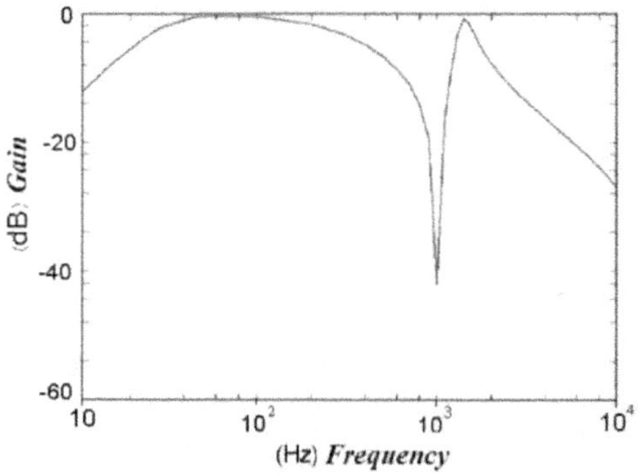


Figure 3-18- Frequency response of output 2

### 3.3.3 Proposed Method

A flowchart for the proposed method is shown in Figure 3-19. The first step is the calculation of $I_0$ either from the output of CT or CVT. The next step is the $I_0$ detection by wavelet transform algorithm. Once $I_0$ is detected, the time can be stamped by GPS clock.

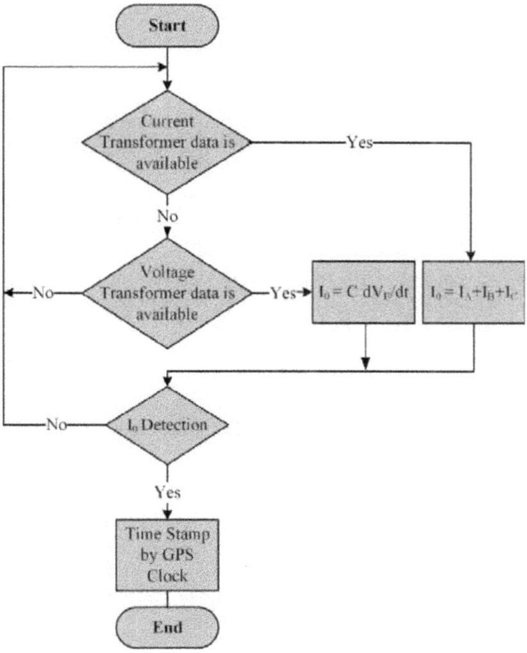


Figure 3-19- Proposed zero-sequence detection flowchart

When a ground fault occurs on a transmission line, impulse signal or travelling wave of $I_0$ is generated. Based on the proposed method, the wave can be detected in the both ends of the line. Detection time has to be stamped with high accuracy of time synchronization system (Figure 3-20).

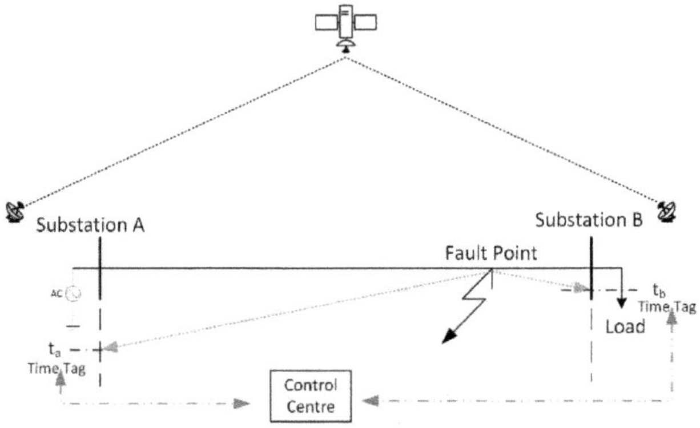


Figure 3-20- Time synchronization in different substations

The estimated fault distance from substation "A" can be calculated from Eq. 3.17.

$$L_F = \frac{L}{2} + (t_1 - t_2)v \quad \text{Eq. 3.17}$$

Where L is the line length, $t_1$ is the zero sequence component detection in substation "A", $t_2$ is the zero component detection in the substation "B", and $\upsilon$ is the speed of the electrical wave in the transmission line which is approximately 0.83 of the light speed in our simulated transmission line.

### 3.3.4 Lab Setup and Evaluation

Figure 3-21 illustrates the experimental setup for testing of fault location function. In this setup, we can simulate a fault in PSCAD software. The transient signals of simulated fault can be saved with the COMTRADE format [132] where a secondary test kit (CMC-256) can playback it and generate the real three phases voltage and current of the fault. The three phase output signals of the CMC256-6 are connected to a transmission line protection relay. In the setup, we applied the SEL-421 relay which is a line protection relay. The relay is synchronized with the GPS clock. In this setup, the SEL-2407 is utilized as a satellite synchronized clock.

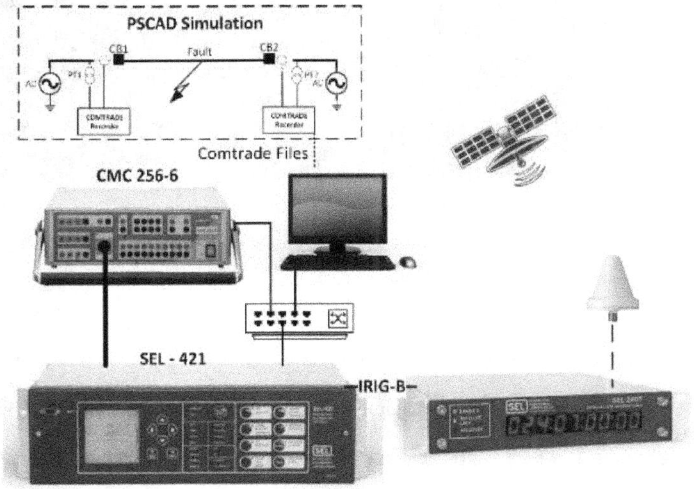


Figure 3-21- Lab setup

Figure 3-22 shows an event viewer of a BG fault which shows the time of $I_0$ detection and fault detection. Similar observations can be archived for the other ground faults.

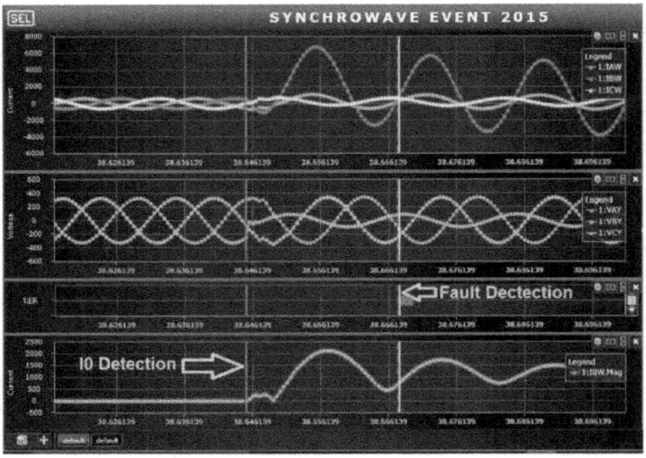


Figure 3-22- $I_0$ detection and fault detection for a BG fault

The accuracy of time is very important factor which depends on the sampling frequency of measuring system as well as time synchronization. Figure 3-23 shows the time difference between two samples are 0.0005 seconds (2000 samples per second) in a relay IED where the higher sampling frequency is not economic. Thereafter, TW-based fault location has to be a standalone device which is not convenient in application.

With the proposed scheme, $I_0$ detection can be applied as a digital value instead of three phase analog signals. Therefore, it can be monitored as an input or output signal with a higher sampling frequency resulting in a higher accuracy.

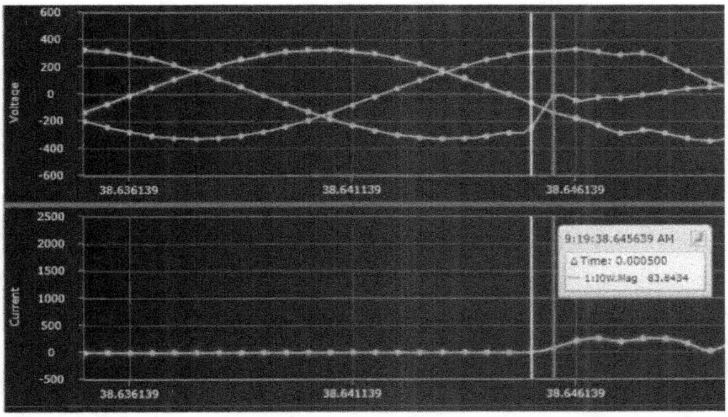


Figure 3-23- Accuracy of $I_0$ detection

The PSCAD modeller with the following fault features is used to evaluate the proposed scheme.

- Sampling frequency: 50 us
- Fault: ABG fault at 0.22
- Location: 25 km far from substation A

$I_0$ is detected at 0.220048783186 and 0.220399225664 second in substation A and substation B respectively in PSCAD software (Figure 3-24).

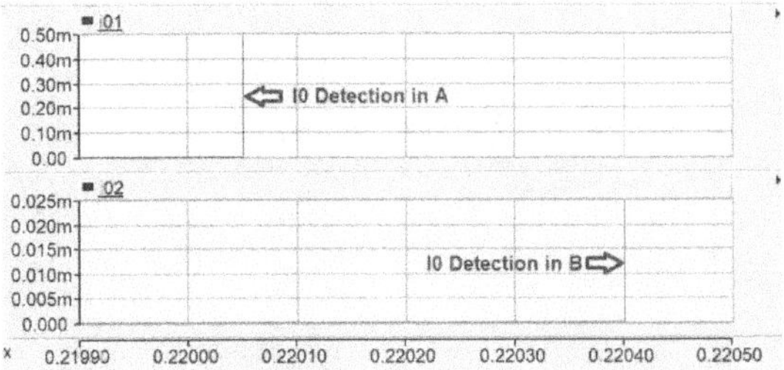


Figure 3-24- $I_0$ detection in substation 1 and 2 for the simulated fault

Based on Eq. 3.17, the estimation of fault location can be obtained as 24.99871855 km far from substation A. Several faults with different location, type and resistance are simulated with the proposed fault detection to analyse the proposed method. The summary of results is listed in Table 3-5. To compare the accuracy of fault location based on different methods, Eq. 3.9 is used.

Table 3-5- comparison of accuracy

| **Fault Type** | **$R_{Fault}$ Ω** | **Location km** | **Fault Location Error (%)** | | |
|---|---|---|---|---|---|
| | | | **With two ends method (experimental Result)** | **Relay Fault Location Function (experimental Result)** | **With the proposed scheme (Simulation Result)** |
| AB-G | 0 | 25 | 5.4667 | 6.3377 | -0.00057 |
| | | 50 | 4.7556 | 5.8222 | 0.194251 |
| | | 100 | 3.3333 | 4.9333 | 0.016754 |
| | | 125 | 2.6222 | 4.6133 | 0.012708 |
| | | 175 | 1.2889 | 2.24 | -0.15742 |
| | | 200 | 0.5333 | 4.4044 | 0.025121 |
| | 50 | 25 | 10.1778 | 188.25 | -5.27913 |
| | | 50 | 4.6667 | 166.8178 | 0.034667 |
| | | 100 | 3.2444 | 139.2133 | 0.031485 |
| | | 125 | 2.5333 | 143.2489 | -0.07568 |
| | | 175 | 1.7333 | 191.87 | -0.08377 |
| | | 200 | 1.0667 | 228.4 | 0.00057 |
| BG | 0 | 25 | 5.1556 | 5.6711 | -0.00057 |
| | | 50 | 4.4444 | 5.9377 | -0.11264 |
| | | 100 | 3.1111 | 6.6488 | -0.02253 |
| | | 125 | 2.2667 | 7.1155 | 0.061811 |
| | | 175 | 0.8889 | 8.2133 | 0.08809 |
| | | 200 | 0.2222 | 8.7911 | 0.00057 |
| | 50 | 25 | 5.1556 | 5.6933 | 0.048533 |
| | | 50 | 4.3556 | 6.6933 | 0.108321 |
| | | 100 | 2.8444 | 8.8933 | -0.11582 |
| | | 125 | 2.1333 | 10.6889 | -0.01675 |
| | | 175 | 1.6889 | 17.1289 | 0.08809 |
| | | 200 | 1.3778 | 23.3778 | 0.00057 |

## 3.4 Conclusion

Recent development in electronic instrument transformer and communication technology provides the new facility for power system functions. This chapter has outlined a new fault location scheme for long transmission lines with the use of the event recorded by two DFRs which is installed in a transmission line. Preliminary analysis of testing the proposed fault locating scheme has shown satisfactory performance and a higher accuracy.

TW-based fault location has been a standalone device which is not convenient in application. The combination of TW-based fault location with impedance-based fault location can reduce the costs, improve convenience, and provide a superior fault location function in a protection relay. To integrate TW-based fault location with impedance-based fault location, this research proposed to apply $I_0$ detection as the initial time of arrival of a ground fault wave, where it can be applied as an input to a relay. A laboratory setup had been developed to discuss the problems of existing protection relays in TW-based fault location and to compare the accuracy of the proposed method with the impedance-based fault location methods.

The proposed scheme for impedance-based fault location can be utilized for other purposes such as protection function which is discussed in the next chapter. The next chapter will discuss the application of communication in transmission line protection functions.

# Chapter 4 Communication based Transmission Line Protection Functions

## 4.1 Introduction

Nowadays, manufacturers are using distance and differential protection functions for short, medium and long transmission lines. Differential protection presents an accurate result for short lines but fails to make correct result for medium and long transmission lines. Recently, communication technology between substations is developed very fast which allows distributed protection and control functions in transmission lines.

This chapter presents the main transmission line protection functions which require communication facility. Thereafter, a new protection scheme is proposed based on distributed measured values in a transmission line. In the proposed scheme, Fiber Optic Current Sensors (FOCS) are placed in different locations of a line. The measured values with IEC 61850-9-2 LE format have been sent to the substations at both the ends of the line where a new protection function is developed based on these distributed measured

---

The content of this chapter is based on the following papers:

1- Saeed Roostaee, Mini S Thomas, and Shabana Mehfuz," Transmission Line Differential Protection Based on Microprocessor Relays" International Journal Series, Archives of Electronics and Electrical Engineering, Vol.2, No.5, Page 14-26, 2016.
2- Saeed Roostaee, Shabana Mehfuz, and Mini S Thomas" Novel Line Protection Scheme based on the Remote IEC 61850 Sample Values in the Line", IEEE Nanotechnology for Instrumentation and Measurement Workshop, IEEE NANOFIM 2017.

values. The performance of distance, differential, and proposed function is evaluated in PSCAD software.

## 4.2 Distance Relay Basics

The distance protection uses a reach measurement of the impedance of the fault loop. The set zones are compared with this measured reach impedance and issue a trip signal to the associated breaker. There are back up zones for external faults which operate with time delay. As the distance relay makes use of an impedance measurement, the loops in which this impedance is measured must be defined. In a three phase system, there are six individual fault loops that cover all short circuit conditions. Three of these are the phase to phase loops ($L_1$-$L_2$, $L_2$-$L_3$ and $L_3$-$L_1$); the other three are the phase to ground loops ($L_1$-G, $L_2$-G and $L_3$-G). Figure 4-1 and Figure 4-2 show phase ground and phase-phase loop respectively [135].

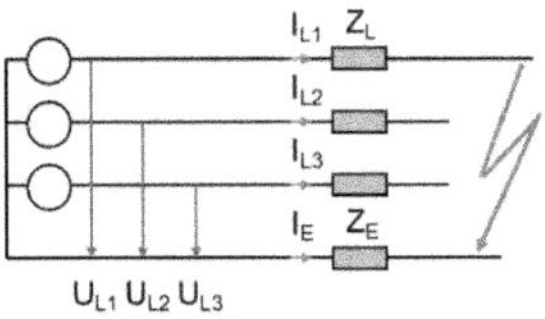


Figure 4-1 - Phase-ground-loop

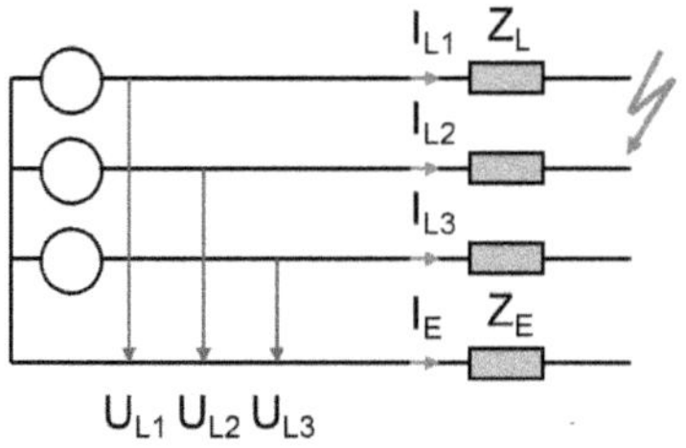


Figure 4-2- Three phase-phase loop

The evaluation of the phase-phase and phase-earth loops does not take place as long as, during single-pole dead time, the affected phase is switched off, to avoid an incorrect measurement with the under defined measured values. The logic block diagram provides the corresponding block signals in each of the state recognition. Figure 4-3 and Figure 4-4 show the block diagram of the phase-earth and phase-phase measuring system correspondingly [135].

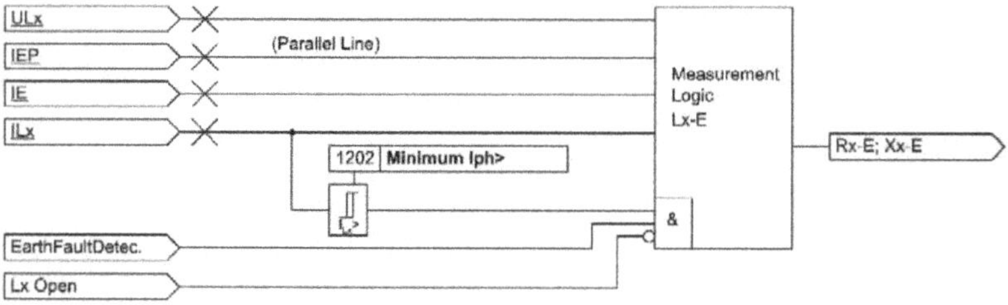


Figure 4-3- Logic of the phase-earth measuring system [135]

The distance relay compares the measured fault loop impedance with its set operating characteristic. The set operating zone setting covers the protected line, in this case, $Z_L$. In practice, some fault resistance ($R_F$) is also present at the fault location.

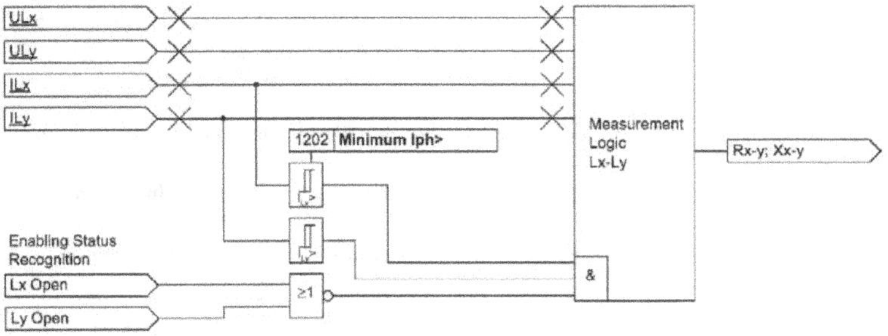


Figure 4-4- Logic of the phase-phase measuring system [135]

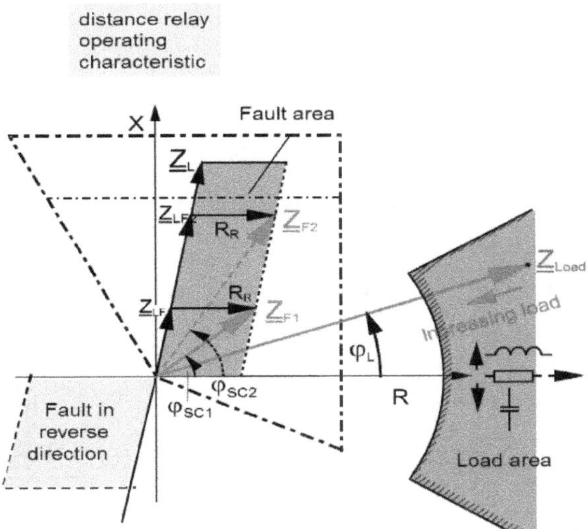


Figure 4-5- Load and short-circuit impedances

This resistance causes a displacement of the loop impedance in the R direction by the amount $R_R$ as shown in Figure 4-5. Therefore, the set zone includes a resistive margin to ensure that the fault resistance does not cause the measured fault impedance to lie outside the characteristic. On long lines with heavy load, the load impedance may encroach on the distance reach setting. The minimum load impedance can be calculated from the minimum operational system voltage and the maximum load current. Apart from this minimum load impedance setting the load encroachment area also has an angle setting which may be derived from the smallest operational power factor.

### 4.2.1 Conventional Implementation of Accelerated Schemes

Figure 4-6 shows the conventional implementation of accelerated transmission line protection scheme using an external communication device. In this implementation, at the sending substation a binary contact of the IED is wired to the tele-protection IED. While in the receiving substation, an output of the tele-protection IED is wired to the receiving IED which the accelerating scheme is used.

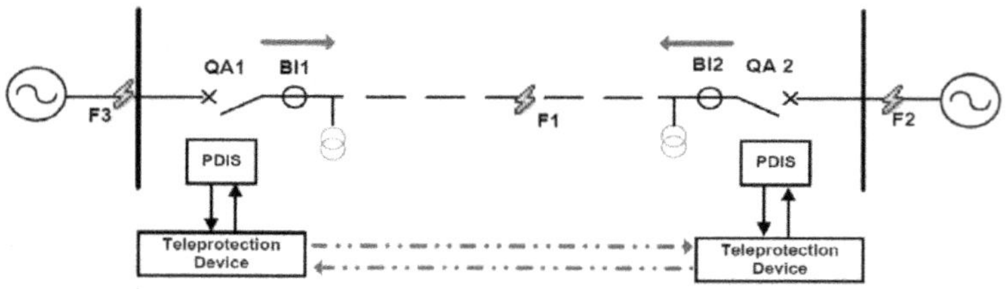


Figure 4-6- Conventional implementation of accelerated scheme using an external communication IED

The availability of communication interfaces in the IEDs allowed the implementation of the accelerated scheme without tele-protection IEDs. In this case, a proprietary protocol is used between the IEDs at the both substations. Figure 4-7 shows the implementation of the accelerated scheme implementation using direct communication between IEDs in different substations.

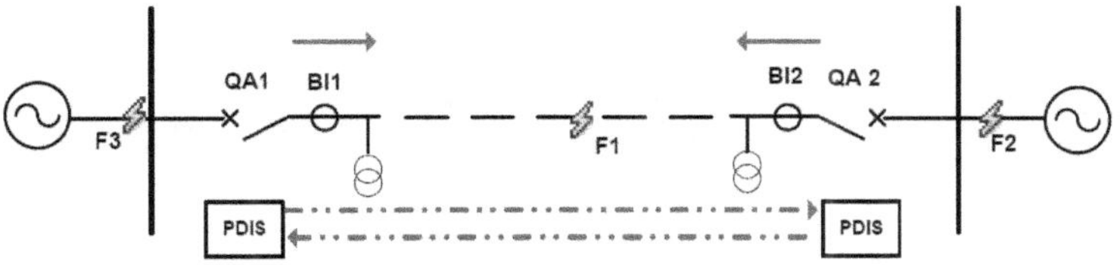


Figure 4-7- Conventional implementation of accelerated scheme using direct relay-to-relay communication

## 4.3 Distance Protection Inter-trip Scheme

In order to provide main and backup protection, power system protection is divided into different zones, based on the reach of the relay. A relay detects a fault in Zone-I and operates instantaneously, while it operates with a delay for faults in subsequent zones (i.e. Zone-II and so on). The trip command for a fault sensed by the distance relay in its Zone-I is based on traditional comparators which compare the impedance with the zone reach impedance setting. An R/X diagram is used to determine the operating characteristic of the distance relay as shown in Figure 4-8. If the impedance measured by the relay is less than a set point value, the relay operates and if it is more than the set point value the relay does not operate. For all values of impedance lying inside the circle in the R-X diagram, the relay operates with the adjusted time delay.

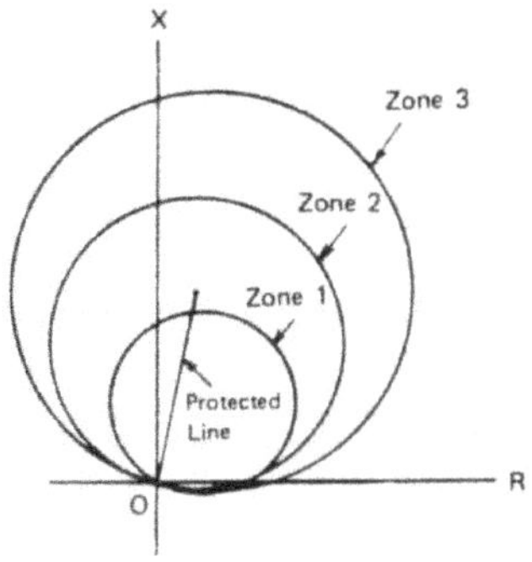


Figure 4-8- R-X diagram of distance relay

In order to increase the fault clearing time, the operating time of the relay can be reduced by employing communication based schemes. The pre-requisite for communication assisted distance relaying scheme is the requirement for a teleportation scheme which transports the relaying signal between the distance relays at two ends. The teleportation scheme for distance relaying can be in three different ways which are discussed in the subsequent subsections.

### 4.3.1 Distance Protection with Permissive Overreach Tele-Protection Scheme

Function diagram of the permissive overreach transfer trip method is shown in Figure 4-9. In this scheme, the distance relay detects a fault in the overreach and sends a signal to the remote end relay via a communication infrastructure. The other distance relay operates only if it receives a true signal from the local end relay. The actors involved in the permissive overreach distance protection scheme are the measuring instrument, sending communication infrastructure, receiving communication infrastructure, circuit breaker. The measuring CT/PT measures the voltage and current of the protected line and is sent to the local relay after data sampling and filtering. When the local relay detects a forward fault, it sends a signal to the remote relay via communication infrastructure. The remote relay also detects a forward fault, and receives a permissive signal from the local end through receiving communication infrastructure; it issues a trip command to the circuit breaker. The logic diagram of this scheme of a Siemens IED is shown in Figure 4-10 [135].

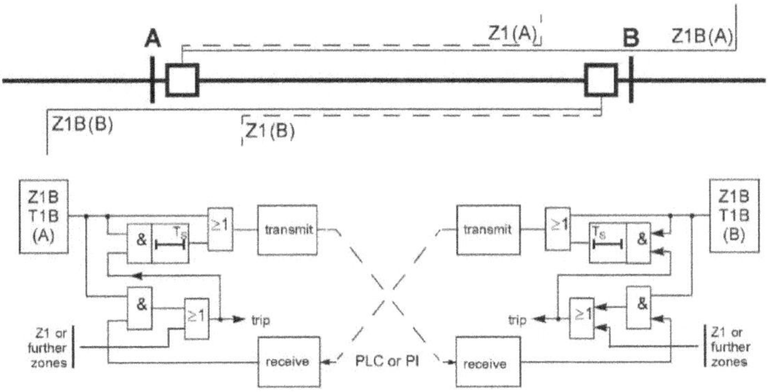


Figure 4-9- Function diagram of the POTT

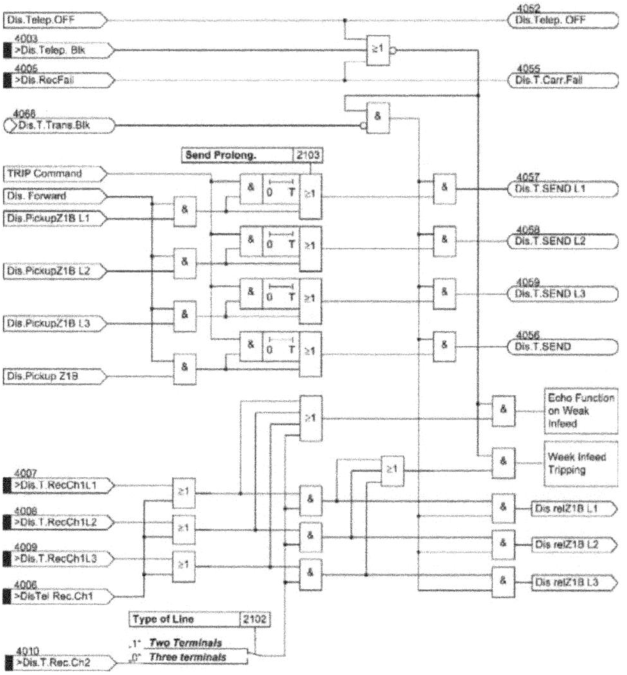


Figure 4-10- Logic diagram of the POTT [135].

### 4.3.2 Distance Line Protection with Blocking Tele-Protection Scheme

Function diagram of the blocking scheme is shown in Figure 4-11. In this scheme, tripping is permitted when no signal has been received and a protection operation has occurred. When a distance relay detects a reverse fault, a blocking signal is sent to the remote end. Whereas if the relay detects a forward fault and does not receive any blocking signal, it sends a trip signal to local circuit breaker to clear the fault. The minimum message size of signal for blocking signal must be one bit which can be increased depending upon the application. Since the blocking signal is meant for tripping, it is necessary that the signal is received fast and the communication channel must be reliable.

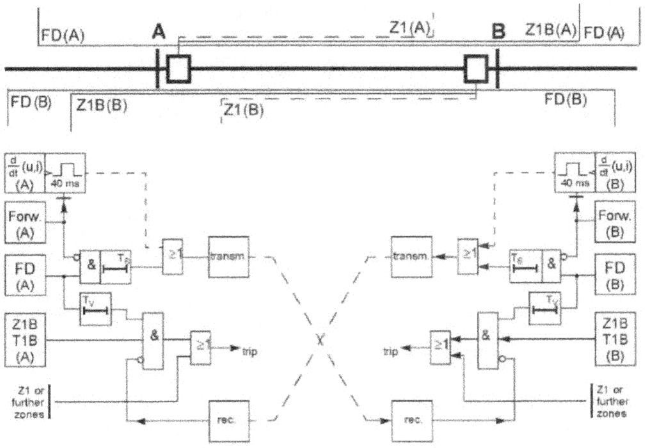


Figure 4-11 Function diagram of the blocking scheme

### 4.3.3 Transfer/Direct Tripping

Function diagram of the direct transfer trip scheme is shown in Figure 4-12. In this scheme, local relay detects a fault and sends a tripping signal to remote the relay to clear the fault. The remote relay trips only if it receives the signal from the local relay. This is also referred to as inter-tripping. Since the received signal for direct tripping no checking of communication channel is involved at the receiving end, a high reliability of the channel is needed.

If the communication assisted accelerated distance protection scheme is through standardized means, there are several advantages such as interoperability, consistency and adaptability to existing substation infrastructure. The subsequent section discusses about adopting IEC 61850 to provide standardized and interoperable communication infrastructure for accelerated distance relaying scheme.

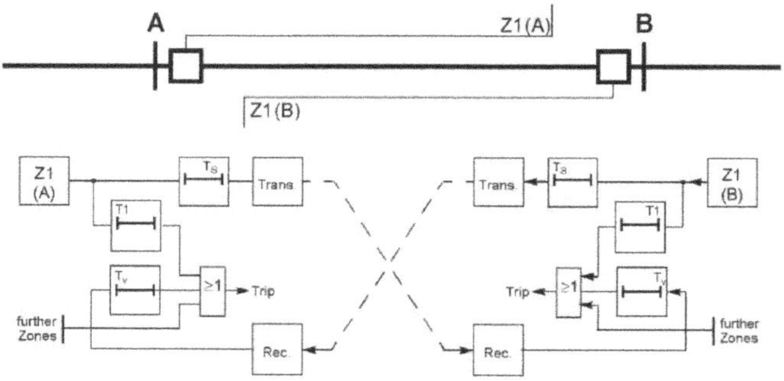


Figure 4-12 Function diagram of the direct transfer trip scheme

## 4.4 Implementation of Accelerated Distance Schemes

Depending on the communication channel and protocol, the implementation of existing accelerated line protection scheme can be either using and external communication device or direct relay-to-relay communication.

### 4.4.1 Implementation of Accelerated Schemes using IEC 61850 GOOSE Messages

The implementation of accelerated scheme using direct exchange of IEC 61850 GOOSE messages between IED in different substations is shown in Figure 4-13. The accelerated scheme can be achieved by publishing GOOSE messages that will be received by the IED which will issue the trip signal to the breaker.

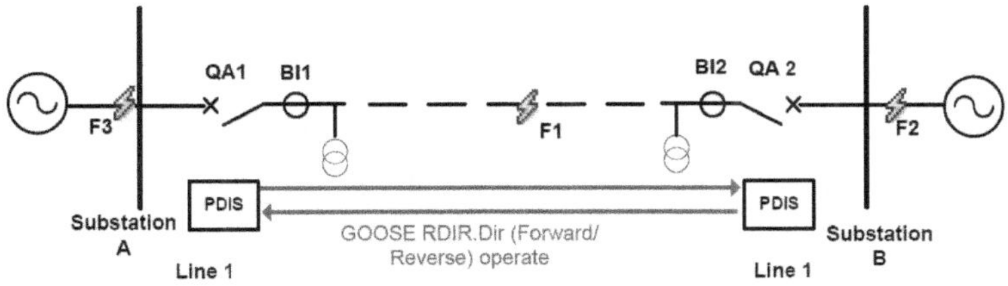


Figure 4-13- Accelerated scheme using GOOSE between IEDs in different substations

## 4.5 Differential Protection

Differential protection is based on the comparison of current. This protection function can be implemented in any electrical device to protect it against internal faults. To do this, one device must be installed at each end of the protected zone. In case of a transmission line, implementation of this is not easy because the data from the both sides of a transmission line have to be compared with each other. Therefore, a communication channel is required. The devices exchange their sampled data via the communication channel and compare them with each other. Figure 4-14 shows the basic of digital differential protection for two ends transmission line. The current at station 1 (S1) is measured by $CT_1$ and the current at station 2 (S2) is measured by $CT_2$. These measured values are sent to a digital differential relay. The current differential protection application is based on Kirchhoff's law. The relay operates when the sum of the current not equal to zero [136]. This rule is valid for the primary system if parallel currents which are produced by line capacitances or magnetizing CT and CT errors be neglected.

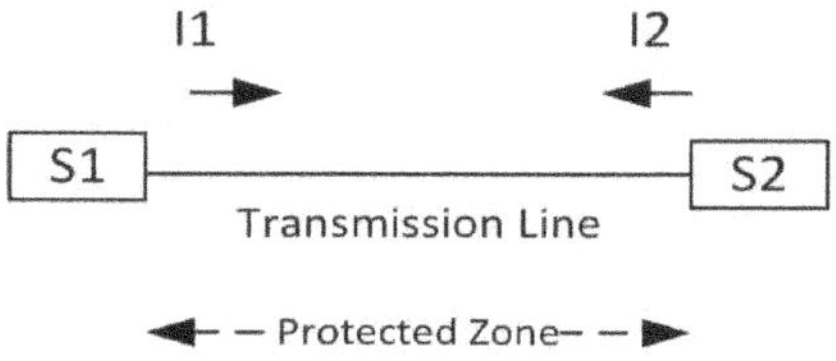


Figure 4-14- Line differential protection

The measuring errors and the transmission line errors can cause deviations of the measured data. Therefore, the total sum of all currents processed in the protected zone in normal time is not exactly equal to zero. With different characteristic area, the differential protection is restrained against these differences. The dual slope percentage biased restraint characteristics which show in Figure 4-15 can be determined by the following settings:

- IS1: The basic differential current setting
- K1: The lower percentage bias setting
- IS2: The bias current threshold setting
- K2: The higher percentage bias setting

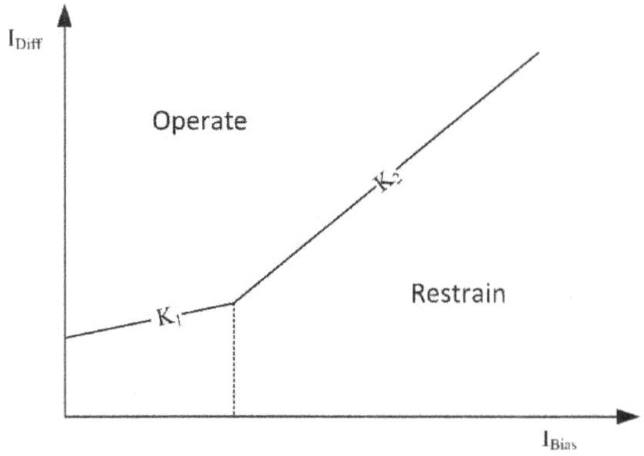


Figure 4-15- Dual slope current differential relay

## 4.6 Performance Comparison of Distance and Differential Scheme

In this section, the performance of a transmission line differential protection function is compared with distance protection. To simulate distance protection, sequence impedance of transmission line is calculated and mapped in R-X diagram. This concept has shown in Figure 4-16. If the measured impedance meets the protected zone in the R-X diagram, distance relay detects and sends the trip signal to the local Circuit Breaker (CB).

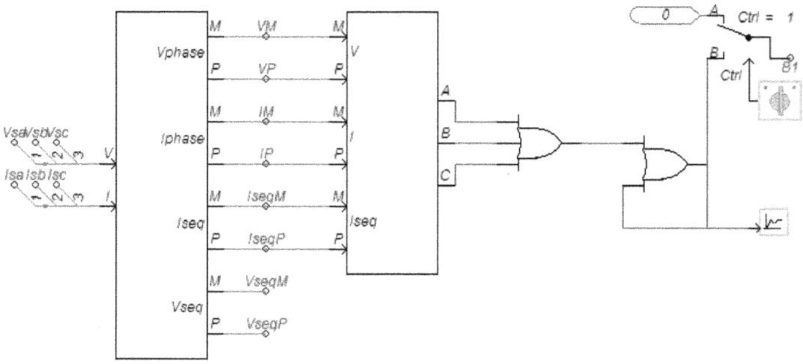


Figure 4-16- Distance protection scheme

To measure the impedance, two blocks are applied. The first block which is shown in Figure 4-17 extracts the fundamental phases and magnitudes of voltage and current based on Fast Fourier Transform (FFT). The outputs of this block are sequence components (positive, negative, and zero) of voltage and current.

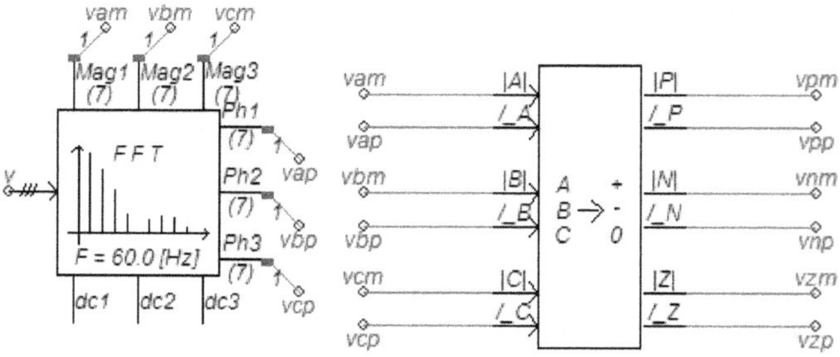


Figure 4-17- Block 1: signal processing of distance relay

The second block which is shown in Figure 4-18 computes the line to ground impedance, and the impedance is mapped in the R-X diagram (Mho Circle).

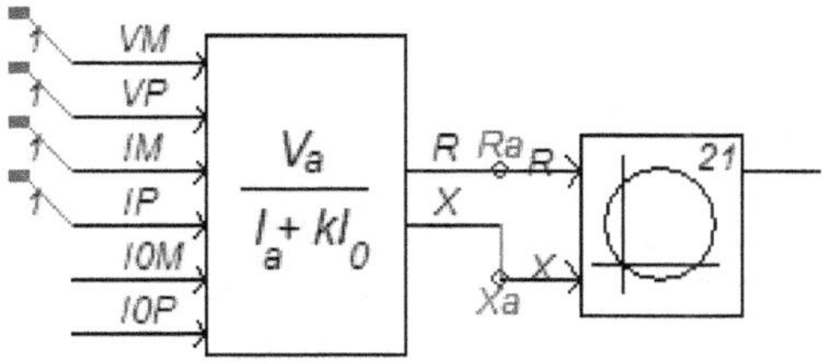


Figure 4-18- Block 2: protection scheme of distance relay

In order to evaluate the performance of distance and differential relays, several faults with different features are simulated. Figure 4-19 shows the trip time fault for an AG fault in 0.2 S without resistance and the location of the fault is in the middle point of the transmission line. Based on simulation result (Figure 4-19), trip time fault for distance and differential trip signal are 0.2167 and 0.2076 correspondingly that one can deduce differential relay is faster than distance relay.

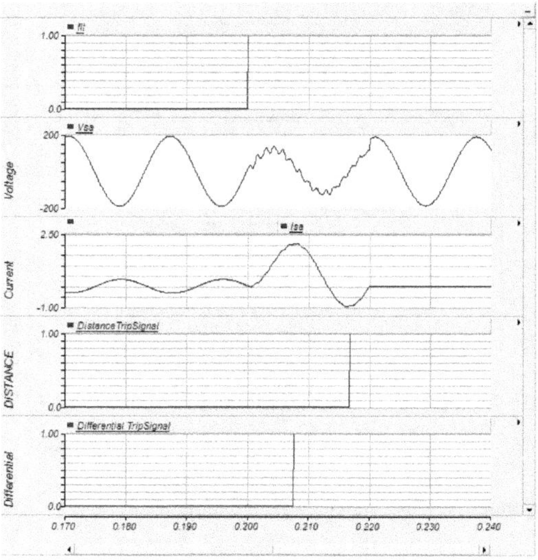


Figure 4-19- Trip signal in distance and differential relays

To compare the fault resistance sensitivity of distance and differential, several faults are simulated. Figure 4-20 shows the simulation result for a 16-ohm resistive fault in the middle of a line. Whereas distance relay could not detect, differential detected the fault. Even differential relay detected a 22-ohm resistive fault that one can deduce differential relay, perform better in resistive faults.

From the simulation result in PSCAD, it is observed the maximum AG resistance fault in the middle transmission line for distance and differential protection is 16 and 22 correspondingly which shows differential relay is more sensitive to detect resistive faults. Whereas distance protection is unable to detect the 22 ohm resisted fault, differential protection is able to detect the same fault (Figure 4-20). Figure 4-21 shows the R-X diagram of distance relay for the 22-ohm resistive fault. In the diagram, impedance characteristics have not met the Mho zone. Hence, the relay would not trip for this fault.

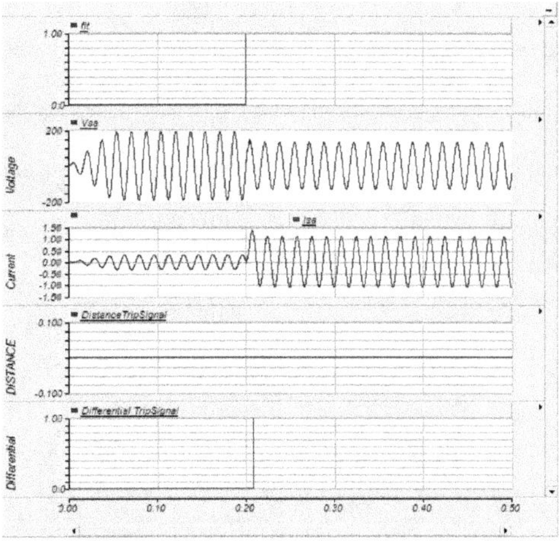

Figure 4-20- Resistive fault in distance and differential

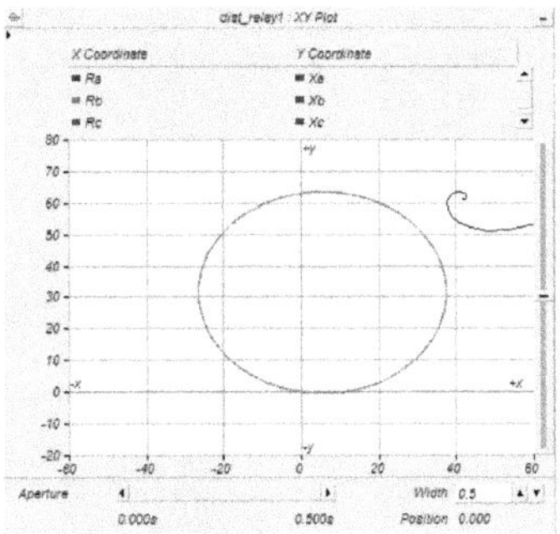

Figure 4-21- R-X diagram

## 4.7 Proposed Protection Scheme

Generally, the line protection functions use the measured values at the both ends of a line. In this thesis, we propose to apply the distributed current of the line as well as the measured values in substations. A similar scheme is also presented in the last chapter for fault location. Figure 4-22 shows the proposed scheme, where two groups of current sensors are installed in the transposed points of a long transmission line. These sensors can send the measured values to the stations through a communication channel.

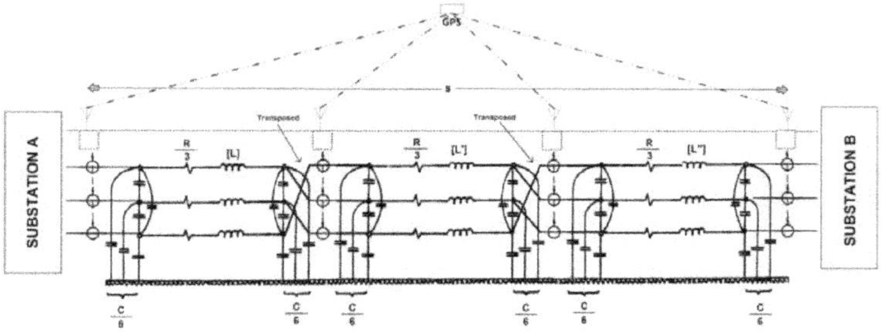


Figure 4-22- Distributed current sensors in transmission line

### 4.7.1 Feasibility of the Proposed Scheme

Fibre-optic current sensors (FOCS) are now available. They can be combined in a light-weight and compact single phase unit with an optical metering unit (OMU). The reduced size and increased accuracy of this optical technology make the design particularly well suited to utilize in the proposed scheme. As the transducers have small size and light-weight, they can be mounted between two insulator strings in a tower.

The OMU is a light device and it can be mounted in a tower. Under these conditions, low-cost fault location has become feasible.

### 4.7.2 Protection Function for the Proposed Scheme

A new protection function should be developed for the proposed scheme. This function can apply the measured value at both the ends of a line as well as the measured values in the line. We have developed a new differential protection based on Kirchhoff's law for the proposed schema. In the other words, the internal current should be equal to the external current. In the designed function, the $I_{a1}$ and $I_{a2}$ are measured by optical sensors which are installed in the substation and the transposed points accordingly, but charging current should be calculated by the following equation.

$$I_{aDiffD1} = I_{X1} + I_{X2} \qquad \text{Eq. 4.1}$$

Where:

$$I_{X1} = \frac{V_a}{X_c}$$

$$I_{X2} = \frac{V_a - I_L((\frac{R}{3}) - L)}{X_c} \qquad \text{Eq. 4.2}$$

Where $I_{aDiffD1}$ is the charging current of section one, $I_{X1}$ is the charging current in substation 1 and $I_{X2}$ is the charging current in transposed point at the end of section one, $V_a$ is the measured voltage in station 1, R is the resistance of the line, $X_c$ is the capacitance of the line, L is inductance of the line and $I_L$ is line current.

Figure 4-23 shows the block diagram for the designed function. In normal condition, the difference of measured currents between a substation and transposed point should be less than the charging current of the line. In a fault condition, differences of $I_{a1}$ and $I_{a2}$ are higher than $I_{aDiff1}$ that the relay detects the fault and issues a trip signal.

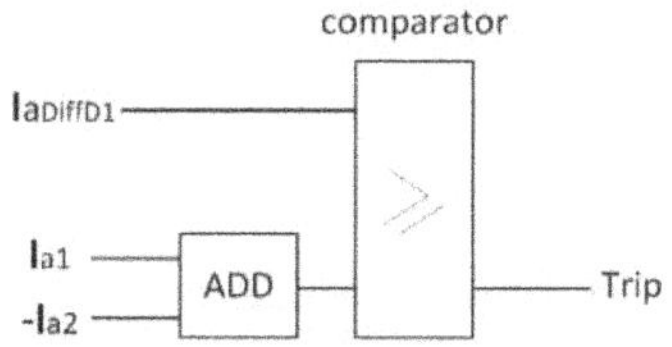


Figure 4-23- Block diagram of the proposed function

The line in section one is not transposed. In the other words, the inductance of each phase is not same. So, the matrix L ([L]) should be applied. Following is the calculation of matrix L:

$$[L]\begin{bmatrix} L_{11} & L_{12} & L_{13} \\ L_{21} & L_{22} & L_{23} \\ L_{31} & L_{32} & L_{33} \end{bmatrix} =$$

$$\begin{bmatrix} \frac{S.Ln\left[\frac{1}{GMR}\right]\mu_0}{6\pi} & \frac{S.Ln\left[\frac{1}{D_{AB}}\right]\mu_0}{6\pi} & \frac{S.Ln\left[\frac{1}{D_{AC}}\right]\mu_0}{6\pi} \\ \frac{S.Ln\left[\frac{1}{D_{BA}}\right]\mu_0}{6\pi} & \frac{S.Ln\left[\frac{1}{GMR}\right]\mu_0}{6\pi} & \frac{S.Ln\left[\frac{1}{D_{CA}}\right]\mu_0}{6\pi} \\ \frac{S.Ln\left[\frac{1}{D_{CA}}\right]\mu_0}{6\pi} & \frac{S.Ln\left[\frac{1}{D_{CB}}\right]\mu_0}{6\pi} & \frac{S.Ln\left[\frac{1}{GMR}\right]\mu_0}{6\pi} \end{bmatrix}$$

Where S is the line length, $D_{AB}$ is the distance between phase A and phase B in the tower, $D_{AC}$ is the distance between phase A and phase C, $D_{CB}$ is the distance between phase C and phase B in the tower, and GMR is Geometrical Mean Radius. Eq. 4.3 shows the simplest form of Eq. 4.1 which can be implemented into a protection relay. The simulation result of a relay with the developed protection function is presented in the next section.

$$\begin{bmatrix} I_{aDiffD1} \\ I_{bDiffD1} \\ I_{cDiffD1} \end{bmatrix} = \begin{bmatrix} -\frac{1}{9}CF\pi(\mathrm{I}_{a1}R-3V_a+3f(\mathrm{I}_{a1}Ln\left[\frac{1}{GMR}\right]+I_bLn\left[\frac{1}{D_{AB}}\right]+I_{C1}Ln\left[\frac{1}{D_{AC}}\right])\mu_0) \\ -\frac{1}{9}CF\pi(\mathrm{I}_{b1}R-3V_b+3f(\mathrm{I}_{b1}Ln\left[\frac{1}{GMR}\right]+I_{a1}Ln\left[\frac{1}{D_{BA}}\right]+I_{C1}Ln\left[\frac{1}{D_{CA}}\right])\mu_0) \\ -\frac{1}{9}CF\pi(\mathrm{I}_{c1}R-3V_c+3f(\mathrm{I}_{c1}Ln\left[\frac{1}{GMR}\right]+I_{a1}Ln\left[\frac{1}{D_{CA}}\right]+I_{b1}Ln\left[\frac{1}{D_{CB}}\right])\mu_0) \end{bmatrix} \quad \text{Eq. 4.3}$$

### 4.7.3 Simulation of the Proposed Scheme

The feasibility of the proposed scheme is tested in PSCAD software. Figure 4-24 shows the simulation of 230 kV transmission line. The length of the transmission line is 300 Km which is divided into three sections of 100 Km based on the proposed scheme. The current sensors are installed at 100 and 200 Km far from the substations. In the other words, they are installed in the transposed points of the transmission line. Therefore, the line is divided into three sections with 100 Km for each. Section one, between substation 1 and 100 km far from the station 1; section two, from 100 to 200 Km of the line; and section 3, from 200 Km to substation 2.

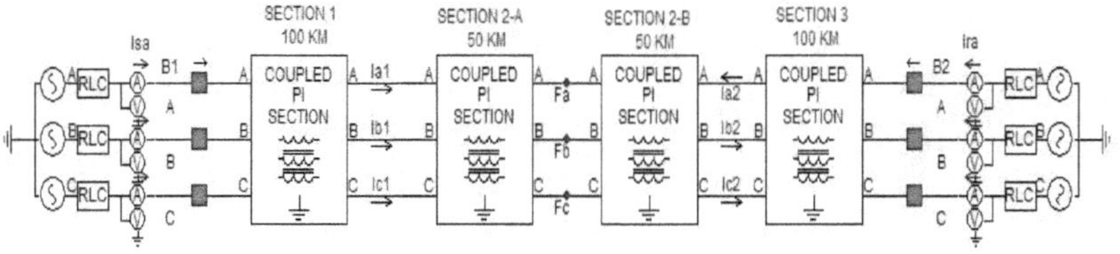


Figure 4-24- Protection function diagram

In section 2, the relay should trip if;

$$\begin{bmatrix} I_{a1} - I_{a2} \\ I_{b1} - I_{b2} \\ I_{c1} - I_{c2} \end{bmatrix} > \begin{bmatrix} I_{aDiffD1} \\ I_{bDiffD1} \\ I_{cDiffD1} \end{bmatrix}$$

To compare the performance of the proposed scheme with the traditional differential protection function, an AG fault in 0.2 seconds with 23-ohm resistance is activated in the middle of the line. Figure 4-25 and Figure 4-26 show the results for the fault in the traditional differential and the proposed scheme respectively. Similar observations have been achieved for several types of fault in the different location which show the proposed differential function is more sensitive in resistive faults.

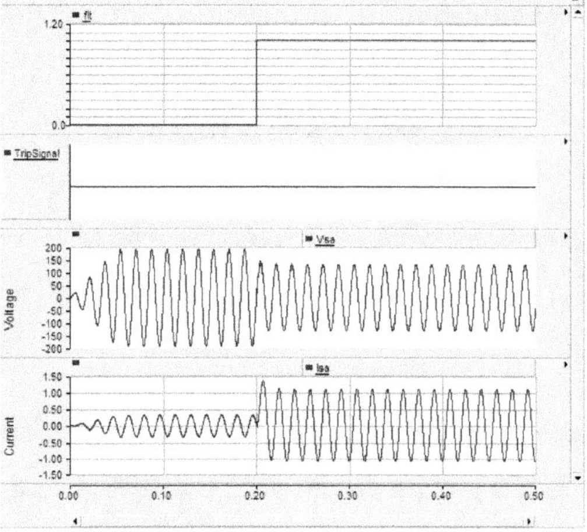


Figure 4-25- Simulation result for traditional differential

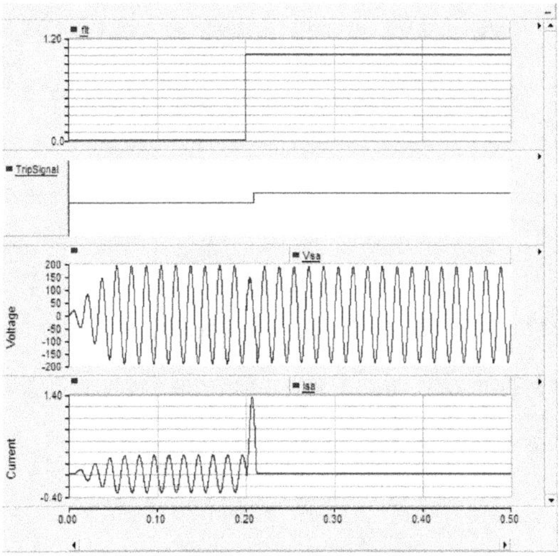


Figure 4-26- Simulating result for proposed scheme

## 4.8 Conclusion

Any mal-operation of protection functions in transmission lines may cause blackout in the whole system. Distance and differential protection functions have been implemented to protect a transmission line for many years. Simulation results show that differential protection function is more sensitive than distance protection in a transmission line.

Communication technology has been developed very fast and smart devices like IEDs are using IEC 61850 for communication. Also, the second edition of IEC 61850 standardizes communication between substations. This chapter proposed a new scheme based on distributed current values in a transmission line which can be available with communication and IEC 61850 at both the ends of a line. A new protection function based on the distributed current value in the line is developed and the feasibility of the proposed scheme is tested using PSCAD modeller. Simulation results have shown that the proposed scheme is more accurate than existing differential protection.

The communication based line protection functions are discussed in this chapter. As the IEC 61850 standard is the latest and the most powerful communication protocol in power system, next chapter will discuss the implementation of existing hardwired functions with IEC 61850 standard.

# Chapter 5 IEC 61850 GOOSE based Interlocking and Distance Protection Inter-trip schemes

## 5.1 Introduction

IEC 61850 is an international standard for communication in substation automation system. The main purpose of the standard is to bring new common communication protocol to the power system automation. The standard is flexible enough to adapt to the future challenges.

IEC 61850 GOOSE based interlocking and assisted distance protection functions are presented in this chapter. Experimental validation for IEC 61850 GOOSE for these two

---

The content of this chapter is based on the following papers:

1- Saeed Roostaee, Mini S Thomas, and Shabana Mehfuz, "Practical Consideration of IEC61850 in Interlocking Functions in Substation Automation System ", 11th INDIACom; IEEE Conference ID: 40353, (BVICAM), New Delhi (INDIA), 2017
2- MohdAsimAftab, Saeed Roostaee, S. M. Suhail Hussain, Ikbal Ali, Mini S Thomas and Shabana Mehfuz, "Performance Evaluation of IEC 61850 GOOSE based distance protection for inter substation communication", IET Generation, Transmission & Distribution (Under review)
3- Saeed Roostaee, Shabana Mehfuz, Mini S Thomas, "Techno-economic assessment of IEC 61850 based Substation Automation System" International Symposium on Resistive Economy: Action and Implementation, JamiaMillialslamia, New Delhi, March 2017
4- Saeed Roostaee, Mini S Thomas, "Automation Function in substation according to IEC 61850", national conference on electrical engineering,JamiaMillia Islamia, ETEEE2015, Delhi 2015

functions has been presented. Also, the performance evaluation of GOOSE messages under different WAN scenarios is reported.

## 5.2 Functions and Interfaces in IEC 61850

In IEC 61850, functions are modelled into Logical Nodes (LNs). The LNs can be define in different levels include; station, bay and process level. For instance, the LN "CILO" includes logical rules for interlocking between switches. It can be allocated in bay level or station level or both of them (Figure 5-1).

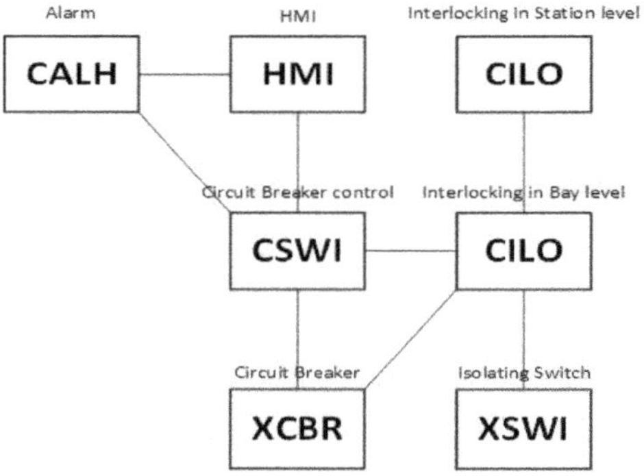


Figure 5-1- LN "CILO" for interlocking in substation

## 5.3 IEC 61850 GOOSE

Until now hardwires are used for P&C functions to share information at bay and station levels which require a large number of cables and consequently the cost of hardware's and maintenance challenged engineers for a long time.

Figure 5-2 demonstrates a traditional protection scheme with hardwire where the protection relays are hardwired for inputs and outputs.

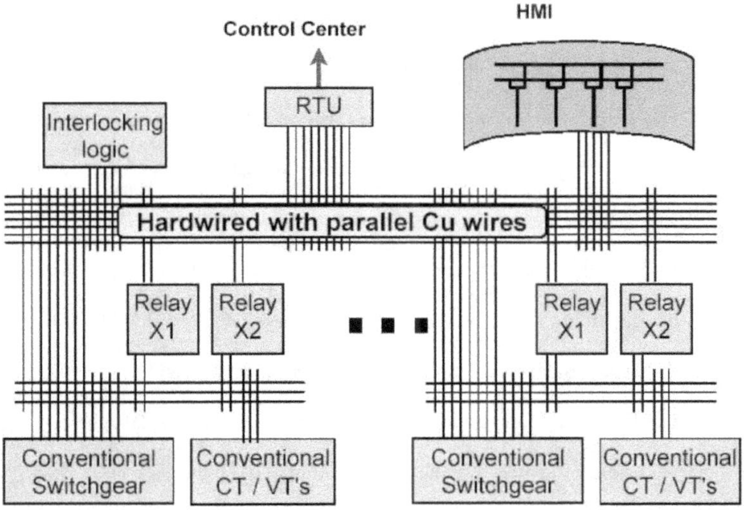


Figure 5-2- Traditional protection scheme with hardwire

A wide range of data set can be exchanged with GOOSE. IEC 61850-8-1 standardized this service which known as IEC 61850 GOOSE. Following sections describe two existing hardwired schemes which are proposed for how their functionality could be implemented using IEC 61850 GOOSE. In the first example, IEC 61850 GOOSE within a substation for interlocking function is considered. And in the second example, IEC 61850 GOOSE between substations for transmission line inter-trip scheme is considered.

## 5.4 IEC 61850 GOOSE based Interlocking

Interlocking function evaluate logical functions to block or operate a switch operation. The closing and opening commands can be issued from local, remote HMI and remote SCADA. To design interlocking function, all of these inputs and outputs must be considered. To implement the logical diagram of this function in a substation, lot of wiring is required. Implementation of IEC 61850 GOOSE based interlocking can reduce the cost of wiring and the maintenance cost. Also, it can improve the reliability by distributed application instead of centralization, and it can be exchanged between devices from the different vendors. All devices which support IEC 61850-8-1 from different vendors can participate in this function. With IEC 61850 complete interlocking in a substation can be performed. In IEC 61850, there is a class for modelling of switch, Double Point Control (DPC), the data class DPC provides a service to implement interlocking function much easier. This data class publish switch information into the network and the subscribers evaluate this data before any switching operation. Also, IEC 61850 provides information report and GOOSE messages to implement interlocking in a substation. Report services use client-server to implement substation interlocking and GOOSE messages use publish-subscribe for bays interlocking. Substation interlocking affects the switching in the whole substation while bay interlocking does it in the bay levels. So each individual bay contains its interlocking functions.

In IEC 61850, there are three different structures to implement the interlocking functions. The first structure is a central interlocking in the station level (Figure 5-3). In this structure; the interlocking functions are concentrated in central IED or server which is placed at station level. In IEC 61850, the position of switches is sent to station level by report control block with client server communication service. In the next structure, interlocking functions are centralized in an IED which is placed at bay level (Figure 5-4). In this scheme, the position of switches sends via report and multicast GOOSE messages to central IED. And finally, the last structure, interlocking functions are distributed in different IED in bay level (Figure 5-5). The position of switches is sent to other IEDs in bay level throughout network.

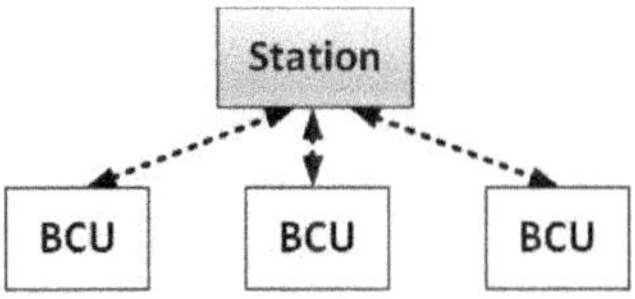

Figure 5-3- Central interlocking in station level

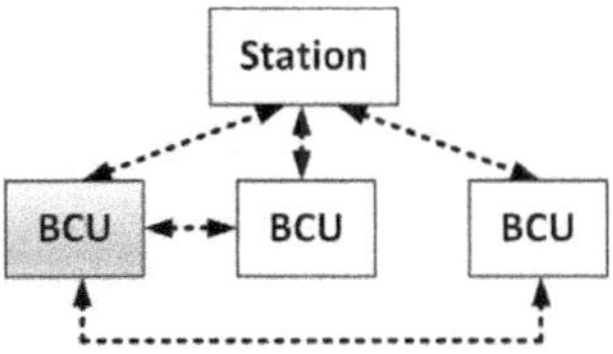

Figure 5-4- Central interlocking in bay level

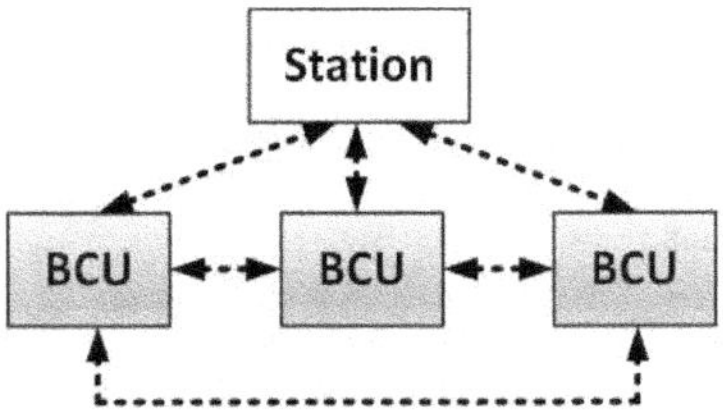


Figure 5-5- Distributed interlocking in bay level

To implement any function based on IEC 61850, some LNs are essential. Minimum required LNs for interlocking function are summarized in Table 5-1.

Table 5-1- Required LNs

| LN | Data | CDC | Attributes | Values | | comment |
|---|---|---|---|---|---|---|
| XSWI | POS | DPC | stVal<br>q<br>t | Intermediate<br>OFF<br>On<br>bad | 0<br>1<br>2<br>3 | Switchgear Status |
| | | | stSeld | True<br>False | 1<br>0 | |
| XCBR | POS | DPC | stVal<br>q<br>t | Intermediate<br>OFF<br>On<br>bad | 0<br>1<br>2<br>3 | Switchgear Status |
| | | | stSeld | True<br>False | 1<br>0 | |
| GGIO | Ind | SPS | StVal<br>q<br>T | True<br>False | 1<br>0 | Grouped the released information |
| CILO | EnaOpn | SPS | StVal<br>q<br>t | True<br>False | 1<br>0 | When the position ON is issued by function |
| | EnaCls | SPS | StVal<br>q<br>t | | | When the position OFF is issued by function |

According to IEC 61850, Data Attribute represents the actual data such as measurements, status, description, and etc. For instance, stVal (status value) indicating actual position of circuit breaker for Data Object type Pos of Logical Node type XCBR. And a Common Data Class (CDC) is a composite set of data attributes, defined by the standard relate to substation applications. For example, a SPS (Single Point Status) can be used to identify the status changes.

### 5.4.1 Interlocking Simulation in PSCAD

To simulate interlocking in PSCAD, we define a new component for Isolating Switch which is shown in Figure 5-6.

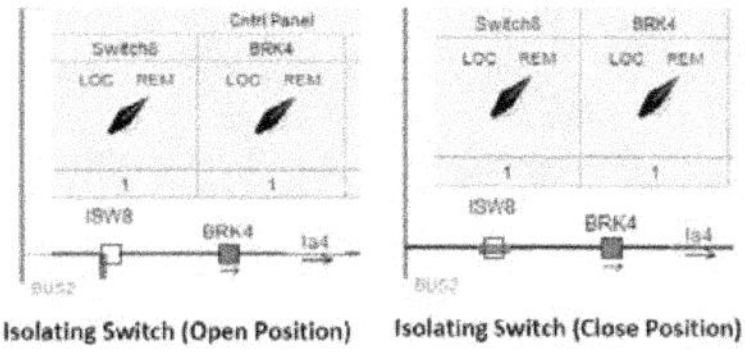


Figure 5-6- Modelling of Isolating switch in PSCAD

One-and-half Circuit Breaker arrangement is simulated in PSCAD to test the interlocking function (Figure 5-7). The logical roles for interlocking and the result of simulation for testing the function from local, remote HMI and remote SCADA are shown in Figure 5-8. The simulation result shows that the designed logic is operating correctly (Figure 5-9).

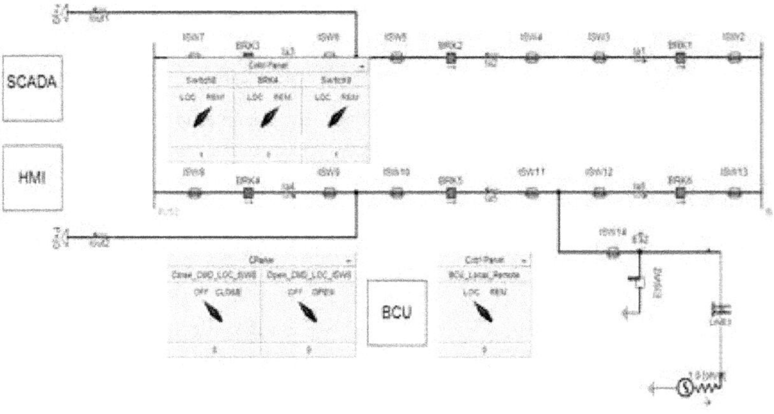


Figure 5-7- One –and – half Circuit Breaker

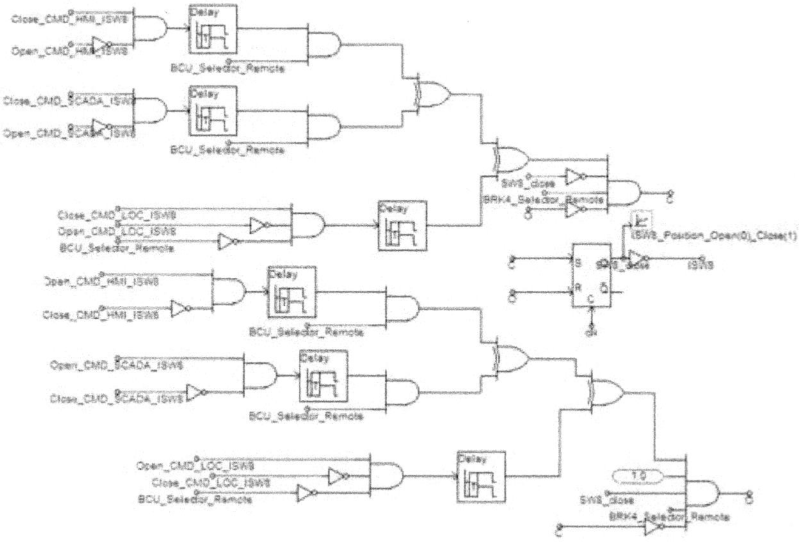


Figure 5-8- Designed Interlocking logic for ISW8

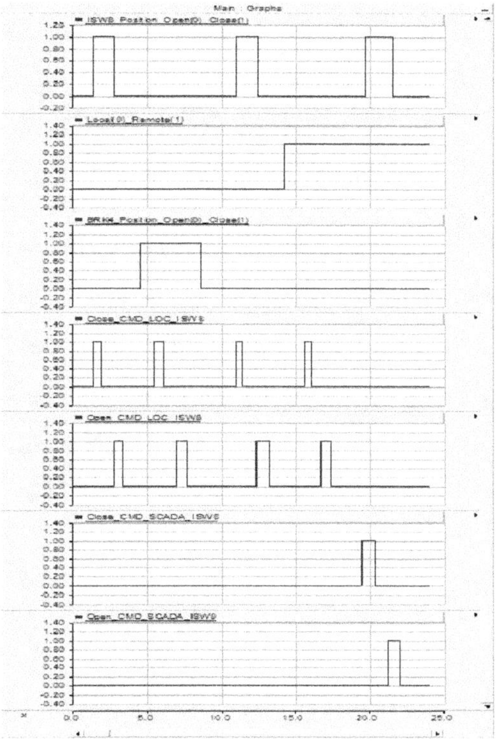

Figure 5-9- Simulation result

### 5.4.2 Implementation in the Lab

This section describes the implementation of interlocking function with IEC 61850 in the laboratory environment at JMI. The first step is to model the information which is required to be published in the format of IEC 61850 GOOSE. For the interlocking application, the information and its type has been summarized in Table 5-2.

Table 5-2- Information to be modelled based on IEC 61850

| Information | State/Command |
|---|---|
| Close_CMD_HMI_ISW8 | State and Command |
| Close_CMD_SCADA_ISW8 | State and Command |
| Close_CMD_LOC_ISW8 | State and Command |
| Open_CMD_HMI_ISW8 | State and Command |
| Open_CMD_SCADA_ISW8 | State and Command |
| Open_CMD_LOC_ISW8 | State and Command |
| BCU_Selector_Remote | State |
| ISW8_Selector_Remote | Command |

The data of switch position is the most important information in the IEC 61850 based interlocking application. For mapping of the switch position, the LN "GGIO" was utilized. This LN uses generic data of switch position's indication and it is published with IEC 61850 GOOSE format. In this set up, GOOSE sender software is utilized for generating GOOSE messages. The configuration of GOOSE sender is shown in Figure 5-10.

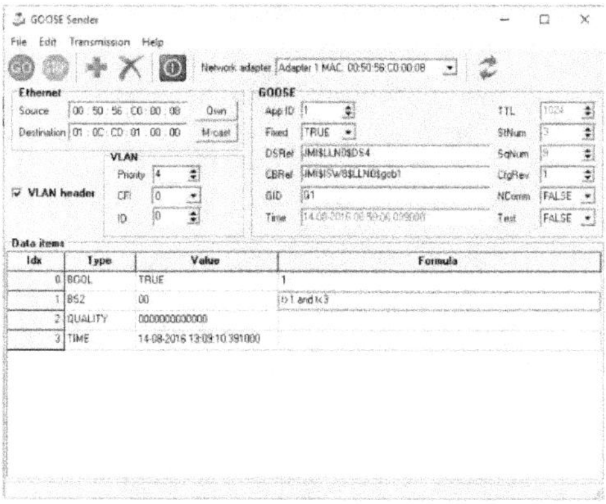


Figure 5-10- Configuration of GOOSE sender

In the setup, the IEDScout is utilized to capture the GOOSE messages. Figure 5-11 shows the subscribed GOOSE message for ISW8 position.

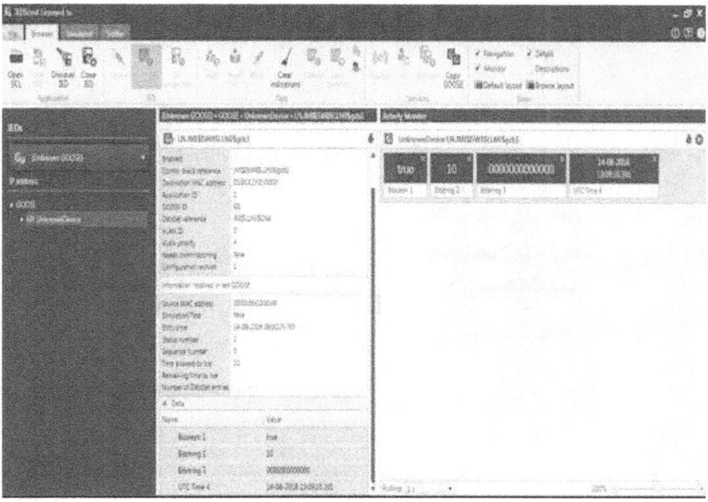


Figure 5-11- The subscribed GOOSE message for ISW8 position

The subscribed GOOSE message is utilized for mapping the data of switch. The 52a and 52b contacts represent the Normally Open (NO) and Normally Close (NC) contacts of isolating switch respectively. Figure 5-12 and Figure 5-13 show the mapping configuration of the subscribed GOOSE with binary outputs and binary input respectively.

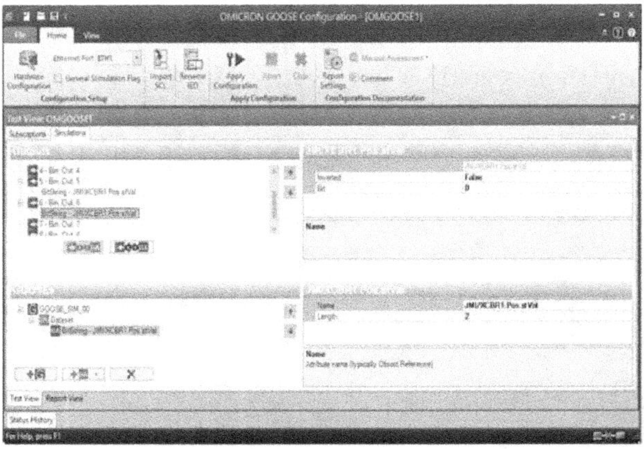

Figure 5-12- GOOSE Configuration for binary outputs

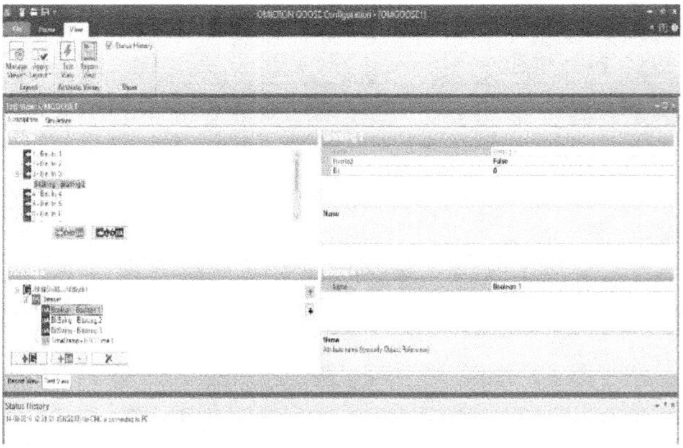

Figure 5-13- GOOSE Configuration for binary input

Figure 5-14 shows the report view of GOOSE module, it shows mapping of subscribed GOOSE into input and outputs.

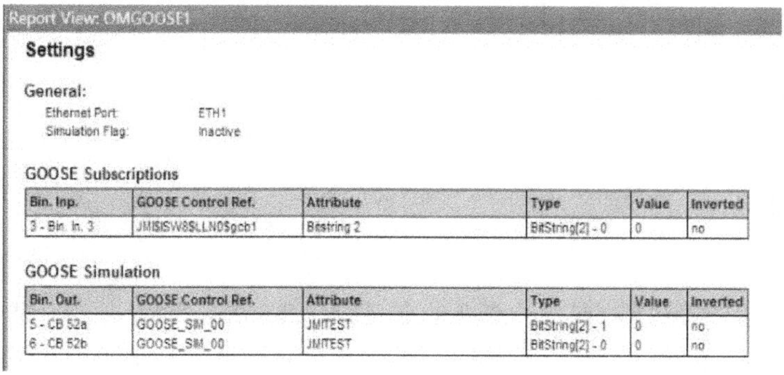


Figure 5-14- Report view

In the setup, we have sent GOOSE signals from a PC, and they are mapped into the binary input/output of CMC 256-6. In order to monitor the network and view the result of test, the Wireshark software is used which is shown in Figure 5-15.

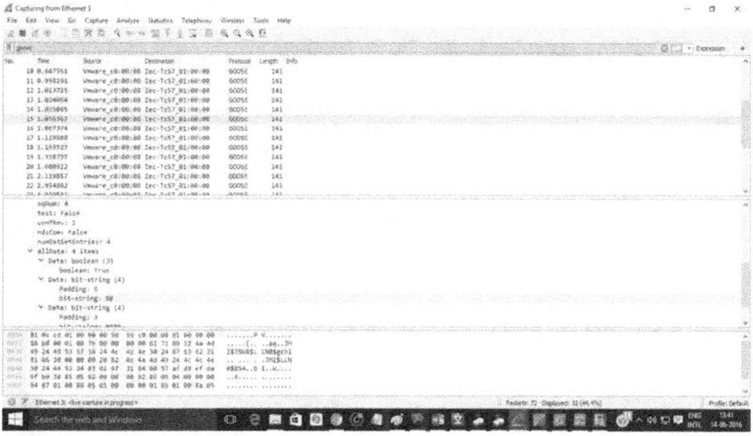

Figure 5-15- Monitoring of network with the Wireshark software

## 5.5 IEC 61850 GOOSE based Distance Protection Inter-Trip Scheme

Distance protection is the most widely used protection technique for transmission line protection due to its enhanced selectivity towards fault, fast response and availability of various relay programming algorithms. However, when faults are detected in the second zone of a relay instantaneous tripping is not possible and is cleared after a coordination time interval. Sometimes this time delay interval may lead to system instability or other substantial effects.

In order to address this issue, an IEC 61850 GOOSE based communication assisted distance protection scheme for protection of transmission line between two substations has been proposed. Experimental validation and performance evaluation for IEC 61850 GOOSE message exchanges for inter substation communication through tunnelling techniques for implementation of accelerated distance protection has been presented. Also, in this section a system-in-the-loop platform has been developed to analyse the ETE delay performances of GOOSE messages under different WAN scenarios is reported.

### 5.5.1 IEC 61850 based Accelerated Distance Protection Relaying

The IEC 61850 based standardized communication assisted distance protection scheme has been implemented by utilizing the GOOSE messages transported between the distantly located substations. Since, GOOSE messages are typically designed for trip signals between IEDs in a substation; they have a fast transfer characteristic. Also, at times of fault, a high amount of data flows in the communication network. The high priority and time criticality of GOOSE messages allow overall performance improvement in the protection scheme in case of faults. The transfer time of a trip and block GOOSE message must be well below 4ms and 20ms [137]. A brief description of GOOSE message is detailed henceforth. GOOSE message are multicast in the network for real time messages and are typically defined for intra-substation communication and thus have no IP/network layer. The Ethertype GOOSE Protocol Data Unit (PDU) along with an expanded Application Protocol Data Unit (APDU) with size of various fields is shown in Figure 5-16. The Ethertype PDU consists of APPID (application identifier) used to distinguish application association, Length field contains the length of Ethertype PDU in bytes, reserved1 and reserved2 are reserved for future applications and Application Protocol Data Unit (APDU). The fields in the GOOSE APDU are encoded using ASN.1 Basic Encoding Rules (BER). ASN.1/BER encoded values always have the format in order Tag, Length followed by Data. The Tag describes the kind of information represented by the field and is of one byte; the length gives the total size of Data in bytes and is of one or three bytes, followed by Data part of the field. The Data may again be in the form of ASN.1/BER encoded value.

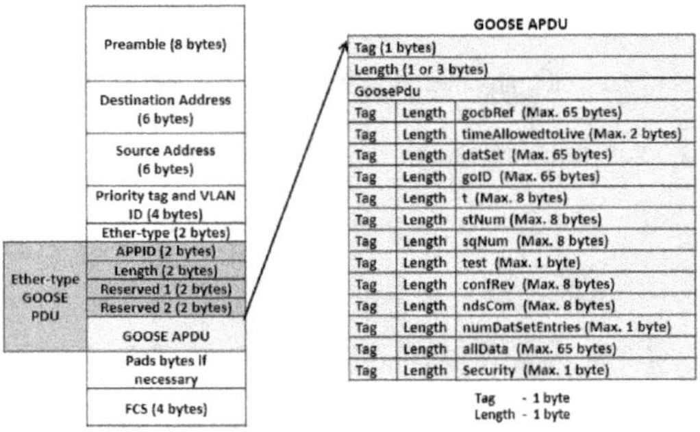


Figure 5-16- Expanded view of GOOSE APDU

The benefits of GOOSE messages as demonstrated in the substation environment makes them also attractive for communications between substations. In order to provide IEC 61850 GOOSE based accelerated distance protection scheme, the GOOSE messages should be transported to the remote substation. In order to send inherent GOOSE message in a WAN, IEC 61850-9-1 recommends the concept of tunnelling. Tunnelling employs encapsulation method to transport data packets from a private network to another network over an IP network (i.e. public network such as Internet).

The router pairs at each end of WAN link forms a virtual tunnel through which the data packets are transported. These routers wrap and unwrap the data packets (here GOOSE packet). Since the inherent GOOSE messages do not possess the third layer of OSI model (IP or network layer), the WAN router subscribes to a particular GOOSE message of interest from its LAN. This GOOSE packet is encapsulated with TCP/IP header and sent over WAN to the receiving router which subscribes it and decapsulates and republishes the GOOSE message onto the receiving LAN. Since tunnelling establishes a virtual link between the two substations in a WAN, the IEDs located in the two substations behave as though they are within the same LAN. This requires the configuration of IEDs in two different substations by exchanging the SCL files via the established tunnel. This has been explained in the subsequent subsection.

### 5.5.2 SCL File Exchange in Tunnelling

In order to provide interoperability and seamless integration in an IEC 61850 process, Substation Configuration Language (SCL) files are required to be exchanged among the IEDs in an IEC 61850 based substation. Configuration of IEDs occurs within a substation by sending the IED Capability Description (ICD) and System Specification Description (SSD) files to the system configuration tool which contains the capability of an IED in form of a blank template and the single line diagram of substation with all the functional requirements respectively. The system configuration tool generates Substation Configuration Description (SCD) file which contains communication configuration of all IEDs in the form they are configured to operate for protection and automation functions in a substation. This SCD file is shared with all substation devices and can be used for multiple applications. The specific IED configuration is then generated from the SCD file containing substation specific address of a particular IED in the form of IED Capability Description (CID) file. Thus, after this process, the complete substation is configured and each IED is aware of locations of all other devices in the substation.

The tunneling process establishes a high bandwidth wide area connection between two different LANs which are geographically distantly located. However, the sender and receiver IED acts as if they are within the same local area network and exchanges the ".icd" files required for IED configuration. This requires the data flow from one IEC 61850 IED in a substation to its appropriate IED in another substation. However, since the power network between the two substations is meshed, this requires that the whole power network information with all substations and IEDs must appear as a single SCL file. For the usual engineering purpose, the concerned IED requires only a part of information of the other system and not the complete network information. Thus, it is appropriate to define a 'system interface exchange description' or SED file. This SED file is exchanged from one substation to another and contains a part of overall information of the system required for a certain purpose.

Since switch is not modelled in SCL, the nearest IED to the switch for each substation is known as the boundary IED for another substation. Referring to Figure 5-17, IED A3 in substation 1 is the boundary IED for substation 2. Similarly, IED B4 in substation 2 is the boundary IED for substation 1. A group of IEDs engineered and configured by the system configuration tool for a specific purpose is known as a project. The engineering control can access the IED for configuration i.e. to change the IED data model rests only with the IED tool. The types of IED engineering rights can be 'fix', for fixing the boundary IED, 'dataflow', for boundary IED data flow addition within the capability of the IED, and 'full', which allows complete engineering of data flows. The SED file is transferred from one substation project to another which augments the

DATA into the data flow. The modified SED file is returned to the originating substation which helps to maintain the same consistent state among the projects.

A case may arise when there are multiple SED files are exported at different points of time. In order to remove this redundancy of multiple versions of SED files, each SED file has identification and a revision index time of modification. This allows identifying the latest SED file version in such cases.

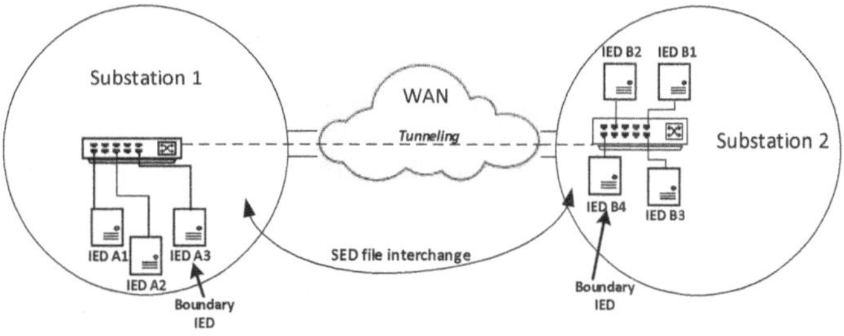


Figure 5-17- SCL file exchanges in tunnelling

### 5.5.3 Laboratory Validation of Problem

A typical power system network was modelled using the PSCAD software. The model consists of a 400 kV, 225 km transmission line, with two substations at each end. Distance relays B1 and B2 are located at end of transmission line along with CT/VT. A double line to ground (LLG-AB-g) fault with a fault resistance of 0.001 Ω is simulated in Zone-I of relay B2 at 25 km from substation 2. The reach of relay B1 is set as 120km, so as the simulated fault lies in Zone-II of relay B2 220 km from it. The fault is added

0.22 s after the simulation begins and is removed after 0.15s. The PSCAD model of transmission line is shown in Figure 5-18.

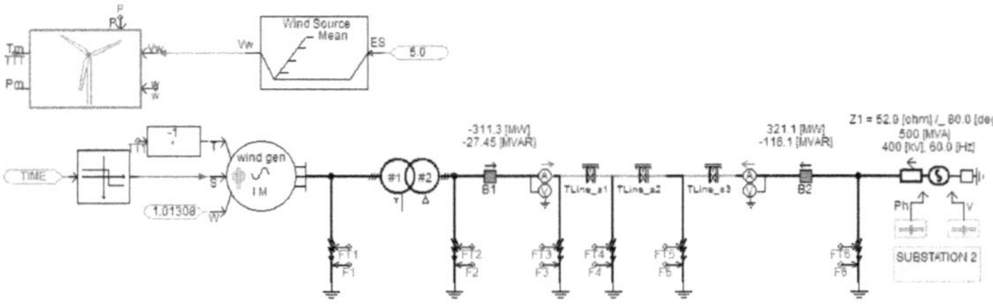


Figure 5-18- PSCAD Model of transmission line

The demonstration and testing of the delay in distance relaying has been carried out on laboratory testbed developed at Substation Automation System (SAS) lab at Jamia Millia Islamia, Delhi (Figure 5-19). This test bed comprises of industrial grade equipment such as IED (SEL 421), Omicron® (CMC 256-6), manageable Ethernet switch, PSCAD simulator as shown in Figure 5-18. The PSCAD simulator is used for development of a power system model. The oculographic and status data of power system transients is stored as COMTRADE file and was playback on the real IEDs using Omicron CMC 256-6 relay testing setup (Figure 5-20).

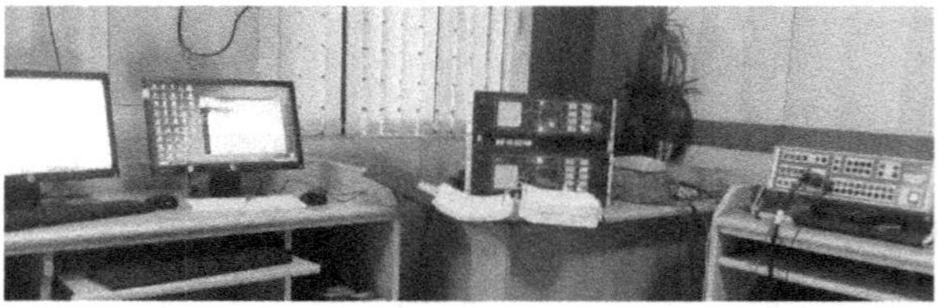

Figure 5-19- Lab setup at Substation Automation at Jamia Millia Islamia

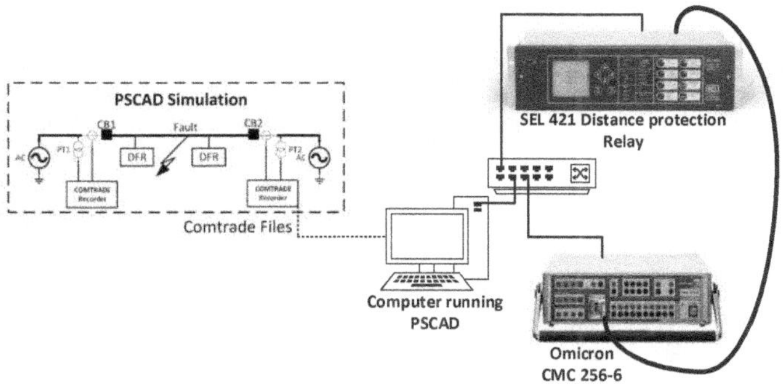


Figure 5-20- Laboratory setup diagram

For the case study with the following fault features, the test is carried out for Zone-I fault for relay B2 and Zone II fault for relay B1.

- Fault Location: 200Km far from Substation 1
- Fault Location: 25Km far from Substation 2
- Fault type: AB-g (A-B-G)
- Fault Start: 0.22 s
- Fault duration: 0.15 s
- Fault Resistance: 0.001 Ohm

The results of recorded three phase voltages (in kV) and three phase currents (in kA) are shown in Figure 5-21 - Figure 5-24. Also, the impedance diagram of distance relays is shown in Figure 5-25. Where the impedance diagrams in distance relay in substation 1 is in the left side, and the impedance diagram of distance relay in substation 2 is shown in the right.

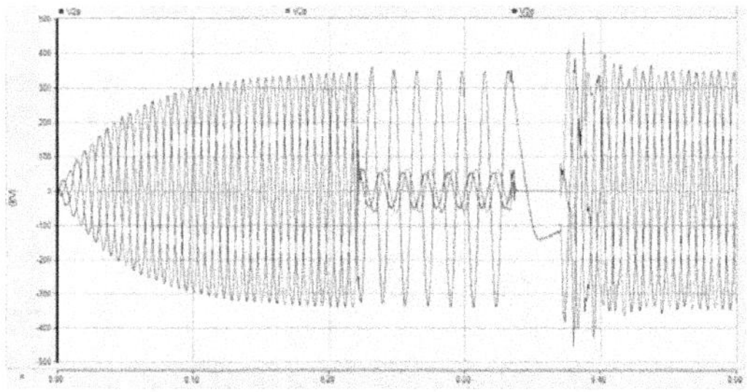

Figure 5-21- Voltage waveform at relay B2

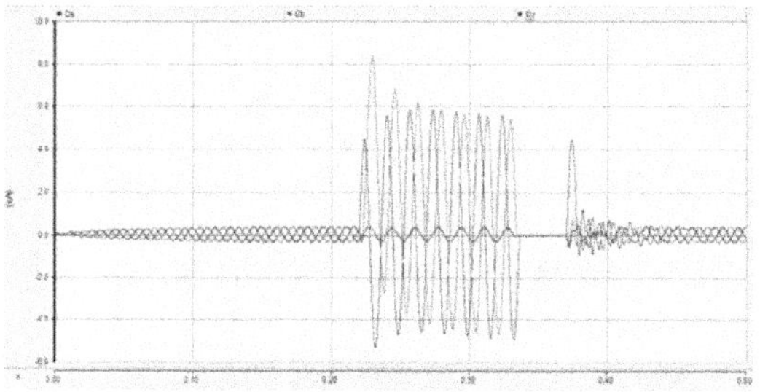

Figure 5-22- Current waveform at relay B2

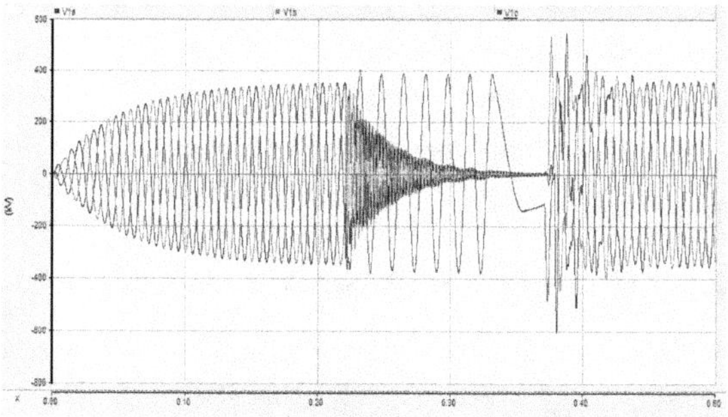

Figure 5-23- Voltage waveform at relay B1

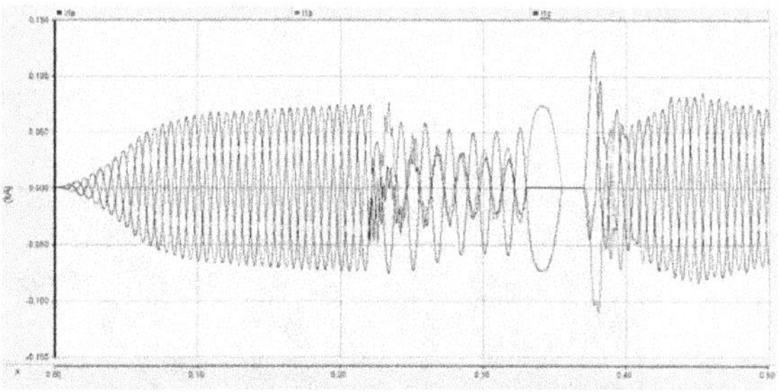

Figure 5-24- Current waveform at relay B1

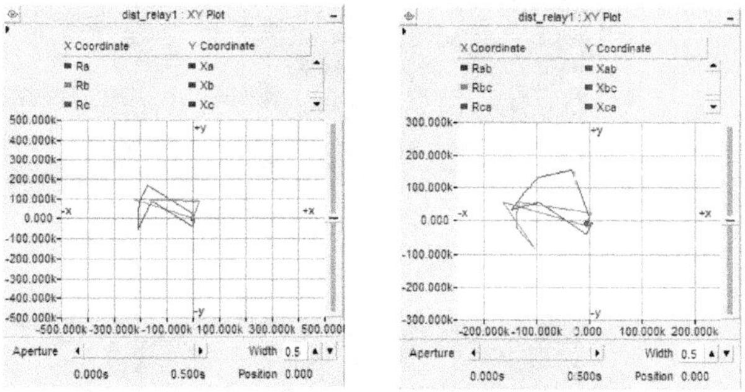


Figure 5-25- Impedance diagram in both substations

The IEDs were tested for distance relay protection. It is found that since relay B2 has a Zone-I fault located at 25 km from the relaying point, the relay trips instantaneously for the AB-g fault. The fault was sensed at around 0.22 sec and the instantaneous voltage and current waveforms are recorded in Figure 5-21 and Figure 5-22 respectively. Figure 5-26 and Figure 5-27 show the HMI of the relay IED (SEL-421) for the substation 1 and substation 2. The experimental results conform to the typical

LLG fault values in the relay B2 (Figure 5-27). However, in case of relay B1, since the fault lies in its Zone-II, the relay B1 fails to trip within the fault duration (Figure 5-26).

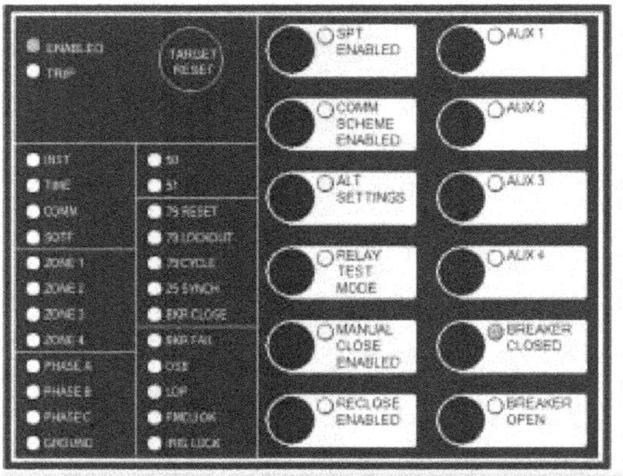


Figure 5-26- Relay in substation B1

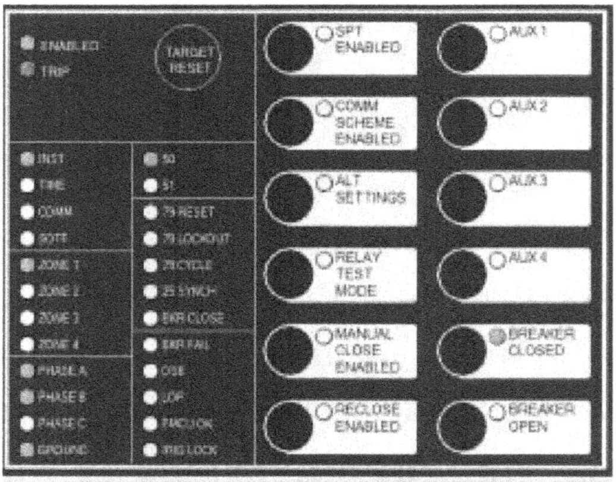


Figure 5-27- Relay in substation B2

### 5.5.4 Performance Evaluation of Inter-Substation Communication Architecture

For evaluating the performance of inter-substation communication network architecture, the transmission line model with substations at each end have been discussed in the previous section, as shown in Figure 5-18, has been considered. The substations are considered to be automated and have their Substation Communication Networks (SCN). The SCN consists of different bays and IEDs namely Protection and Control (P&C) IED, Breaker IED and Merging Unit (MU) IED [93]. The SCN of substation with different bays and IEDs is shown in Figure 5-28. The two SCNs are connected with each other through a wide area network (a meshed network of routers) via gateway routers.

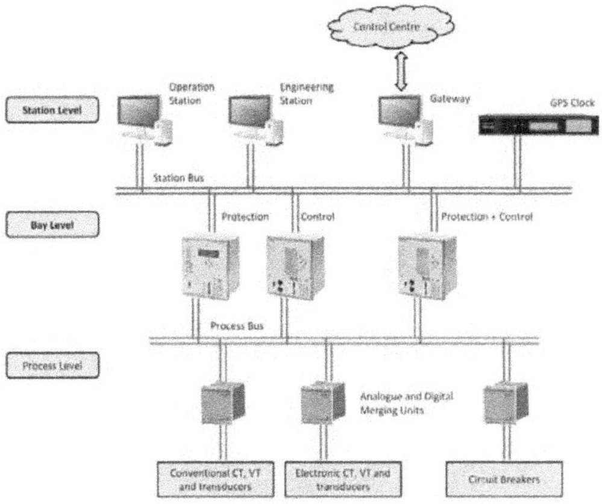


Figure 5-28- Substation Communication Network Architecture

In order to evaluate the inter-substation communication architecture between two substations, a system-in-the-loop (SITL) simulation set up is configured. In the Riverbed Modeler Network simulator software [138], the two SCNs with different bays and IEDs, are simulated as per [93]. And these SCNs are connected through a Wide Area Communication network containing routers spanning over a distance of 225 km. For setting up the SITL simulation the P&C IED in substation 2 and the Breaker IED of the substation 1 are replaced by real IEDs in the simulation with the help of SITL nodes of Riverbed Modeler. The SITL node of Riverbed modeler facilitates an interface which can enable the simulated network to exchange data with a real network. Now, the SITL experimental setup for validation of proposed IEC 61850 GOOSE based accelerated distance protection scheme is as shown in Figure 5-29.

The experimental setup consists of three modules: P&C IED of substation 2, running IEC 61850 GOOSE Sender software by InfoTech [139], a WAN network simulated through Riverbed modeler and Breaker IED of substation 1, running IEC 61850 GOOSE receiver software by InfoTech [139]. The setup is connected in such a way that the P&C IED of substation 2 and Breaker IED of substation 1 are replicated via real devices (computers) acting as GOOSE publisher and subscriber respectively.

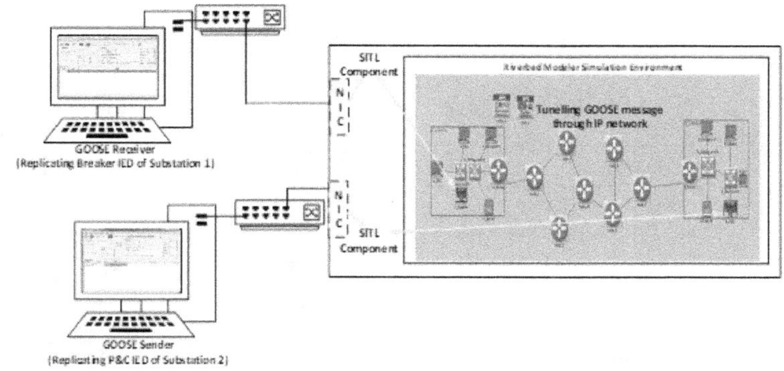


Figure 5-29 Experimental setup for validation of GOOSE assisted distance protection scheme

The modeling of different IEDs is as follows. The MU IED generates the samples of current and voltage from CTs/PTs and converts them into data packet format adhering to IEC 61850- 9-2. It sends the SVs to P&C IEDs and other IEDs of the network in the same LAN. The MU IED is modeled in the Riverbed modeler by selecting appropriate attributes in the *'ethernet_workstation'* node. The P&C IED receives measurement data from the MU IED and based on specific applications generate GOOSE message to the breaker IED. The P&C IED is modeled via real IED running IEC 61850 GOOSE Sender software. The Breaker IED subscribes the GOOSE message from the P&C IED and operates the circuit breaker. The Breaker IED is modeled via real IED running IEC 61850 GOOSE Receiver software.

The traffic i.e. messages exchanged between different IEDs in SCN (intra substation) and inter substation for the proposed accelerated distance protection scheme is given in Table 5-3. In simulation, the router gateways in SCNs are configured with a GRE tunnel. The simulation is run for 5 minutes and the End-to-End (ETE) delays for different messages in SCNs are shown in Table 5-4.

Table 5-3- Traffic in typical Substation Communication Network

| Application | Source IED | IEC 61850 Message type | Packet Size | Destination IED | Frequency (/s) |
|---|---|---|---|---|---|
| **SCN Traffic** | | | | | |
| Sampled Values | MU IED | Type 4 - Raw data message | 126 | P&C IED | 4800 |
| GOOSE | P&C IED | Type 1, 1A | 104 | Breaker IED | - |
| Status Updates | P&C IED, Breaker IED | Type 3 | 200 | Station Sever | 10 |
| File Transfer FTP | P&C IED, Breaker IED, Server | Type 5 - Background traffic | 256 | Server, P&C IED, Breaker IED | 1 |
| **Inter Substation Communication Network Traffic** | | | | | |
| GOOSE Trip Signal | P&C IED | Type 1, 1A | 104 | P&C IED, Breaker IED | 1 |

Table 5-4- Communication traffic in substation along with comparison of average delay for various substation messages

| Application | | IEC 61850 Message type | Packet Size | Delay (average) | |
|---|---|---|---|---|---|
| | | | | Without Priority Tagging | With Priority Tagging |
| Sampled Values | | Type 4 - Raw data message | 126 | 0.242 | 0.210 |
| GOOSE | Intra Substation | Type 1, 1A - GOOSE Trip signal | 104 | 0.58 | 0.36 |
| | Inter Substation | | | 4.35 | 4.12 |
| Status Updates | | Type 3, Status updates | 200 | 0.19 | 0.11 |
| File Transfer FTP | | Type 5 - Background traffic | 256 | 0.15 | 0.13 |

It is assumed that at t=120s a fault occurs in the transmission line which is sensed in Zone 2 of the P&C IED in substation 1. The same fault is detected in Zone 1 by the P&C IED of substation 2. Hence the P&C IED of substation 2 sends a GOOSE message to its breaker (intra substation) as well as sends a GOOSE message to the far end P&C IED at substation 1 via the virtually established tunnel in WAN connecting the two SCNs. This is visualized by setting the P&C IED of Substation 2, i.e. PC running GOOSE sender software, to publish the GOOSE messages at t=120s. These GOOSE messages are subscribed by the Breaker IED in substation 1, i.e. PC running the GOOSE receiver software. The GOOSE packets are transported from SCN 1 to SCN 2 via the tunnel. The sender and receiver transports data packet through the tunnel in same manner as if they are within the same LAN. The captured GOOSE packet at the breaker IED of Substation 1 is shown in Figure 5-30.

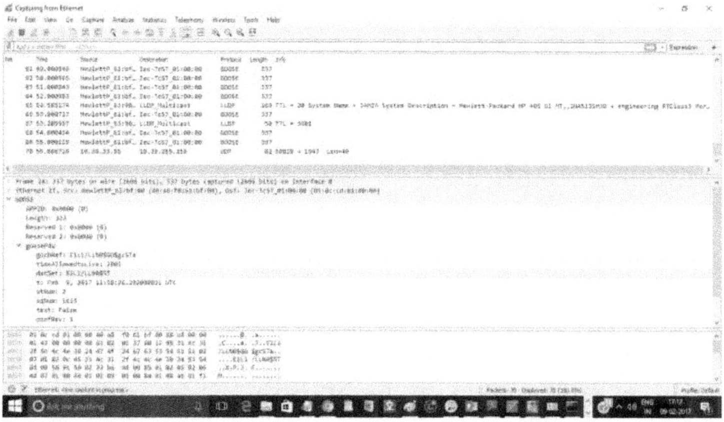

Figure 5-30- Monitoring GOOSE

The major portion of ETE delay for transporting the GOOSE message via tunnel is contributed by encapsulation and decapsulation of the packets at the routers. The average ETE delay for the GOOSE message transmission is computed using Riverbed modeller and is found to be 4.35ms. While, the same GOOSE message is also sent to the Breaker IED of the substation 2 has comparatively less delay, which is due to fact that this breaker IED is in same LAN of SCN of substation 2. Figure 5-31 shows the ETE delays of inter and intra substation GOOSE messages.

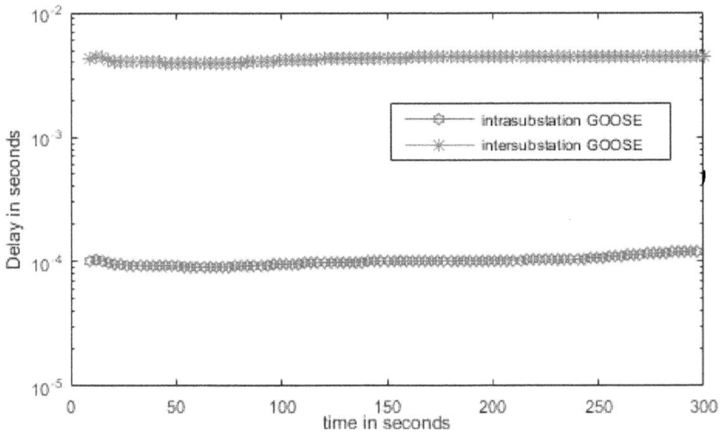


Figure 5-31- E2E delay for intra-substation and inter-substation GOOSE message

The ETE delay performance of wide area inter-substation communication networks under different congestion levels is evaluated. From Figure 5-32 and Table 5-5, it can be noticed that congestion has a considerable impact on the ETE delays in case of links with low bandwidths. However, in the links with high bandwidths, the effect of congestion is minimal. The DS0 links which have bandwidth of 64 kbps are not suitable for this accelerated distance protection scheme for long distances as they result in higher ETE delays. Whereas the links with higher bandwidth in the order of Mbps will have acceptable ETE time delays for message exchanges over longer distances.

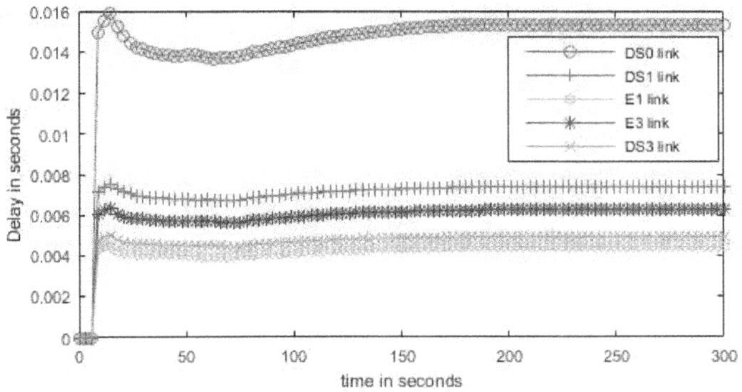


Figure 5-32- Average E2E delay for various types of WAN links

Table 5-5- E2E GOOSE delay for different types of WAN links under different congestion levels

| GOOSE Delay (in ms) | | | | | |
|---|---|---|---|---|---|
| Type of links/ Capacity→ Congestion↓ | DS0 (64 kbps) | DS1 (1.544 Mbps) | DS3 (44.736 Mbps) | E1 (2.048 Mbps) | E3 (34.368 Mbps) |
| No Congestion | 15.23 | 7.31 | 4.35 | 6.15 | 4.82 |
| 25 % congestion | 18.25 | 8.2 | 4.35 | 6.10 | 4.82 |
| 50 % congestion | 26.8 | 8.8 | 4.20 | 6.10 | 4.77 |
| 75 % congestion | 48.5 | 8.8 | 4.13 | 6.10 | 4.72 |

## 5.6 Benefits of IEC 61850 GOOSE

The benefit of the IEC 61850 based functions are summarized as bellow:

- By making use of signalling through process and station busses within a substation, the amount and complexity of wiring can be considerably reduced leading to significant cost savings
- Less wiring also leads not only to fewer installation costs but also less testing and maintenance costs during the lifetime of the substation automation system
- Hardware savings follow due to the fact that the number of I/O in the IEDs can be reduced
- Any subsequent modifications to schemes involving connections between IEDs will also become less expensive as changes to hardwiring will be minimized
- Live testing
- Improve the substation safety by replacing analog circuits with communication links
- Monitoring of the GOOSE messages (copper wired signals are not!) and LAN devices

## 5.7 Conclusion

The implementation of IEC 61850 in substation has many advantages. The standard only defines data information and communication for this application and the implementation of functions is out of the scope of the standard. In this chapter, IEC 61850 GOOSE based interlocking and distance inter-trip are considered. Conventional hardwired substation automation functions can be replaced with communication where hundreds of wires can be replaced with a GOOSE message. An IEC 61850 based interlocking scheme is developed and tested in PSCAD software. To implement the proposed scheme, the required Logical Node is considered and configured in a lab environment.

Fast fault clearance and system restoration help in increasing the power system stability to such an extent that a saving of 1ms time in system restoration helps to improve the power system stability by 15 MW. In order to transport inherent GOOSE messages, with no IP/network layer, for inter-substation communication concept of tunnelling has been used in this thesis. This chapter as presented the performance analysis of accelerated IEC 61850 GOOSE based distance protection scheme according to IEC 61850-9-1. A system-in-the-loop set up for analyzing the ETE delay performances of GOOSE messages for different scenarios has been reported. The performance evaluation carried out in this chapter for different types of WAN links and it has been concluded that links with lower bandwidths are not suitable for long distances as they may result in the high end to end delay.

The implementation of IEC 61850 in power system has many advantages. However, IEC 61850 does not standardize the network topology in SAS. Hence, the customers utilize different structure based on the reliability requirements and economical aspect. Next chapter discusses the different topologies in SAS.

# Chapter 6 Reliability Analysis of SAS based on IEC 61850

## 6.1 Introduction

Demand for reliable electricity is increasing and utilities are offering better quality electricity. In the previous chapters, IEC 61850 based interlocking and distance inter-trip functions are presented. In this chapter, the reliability and availability of IEC 61850 based substation automation system are discussed.

IEC 61850-9-2 and IEC 61850-8-1 specify exchanging GOOSE message and Sample Values through a LAN. These messages are forwarded within a subnet to all its nodes. The standard does not define the structure of the process bus network. In this chapter, the demonstration and testing of different process bus communication topologies have been carried out based on the developed laboratory test bench.

---

The content of this chapter is based on the following papers:

1- Saeed Roostaee, Shabana Mehfuz, Mini S Thomas ," Research and Implementation of the Communication Topologies of Process Bus based on IEC 61850", IEEE Nanotechnology for Instrumentation and Measurement Workshop, IEEE NANOFIM 2017
2- Saeed Roostaee, Shabana Mehfuz, and Mini S Thomas "Reliability Comparison of Various Power Substation Automation based on IEC 61850" International Journal Series in Engineering Science (IJSES) (ISSN: 2455-3328)

## 6.2 Substation Automation Architecture

In IEC 61850, all functions in substation automation system can be divided into three levels. Figure 6-1shows these three levels which are station level, bay level, and process level. Process level is the lowest level. The instrument transformer, sensors and actuators which are necessary to monitor and operation of switchgear are located in this level. The bay level is the middle level which bay controller and IEDs are located in this level. The station level is the upper level of SAS where Human Machine Interface (HMI) and gateways for connection to Network Control Centre (NCC) are located in this level. These levels can be connected to the same IEC 61850 network. However, due to high traffic of Sample Values (SV) at the process level, the process network is separate from station network.

As communication has very important roles in a substation automation system, the reliability of the network should be considered. To do this, Ethernet Switches (ESW) and interfaces should be modelled. As the Ethernet interfaces are part of IEDs and its MTTF is considered in the IED, we consider only communication system. Subsequent sections discuss the five common topologies with different levels of reliability and cost in SAS.

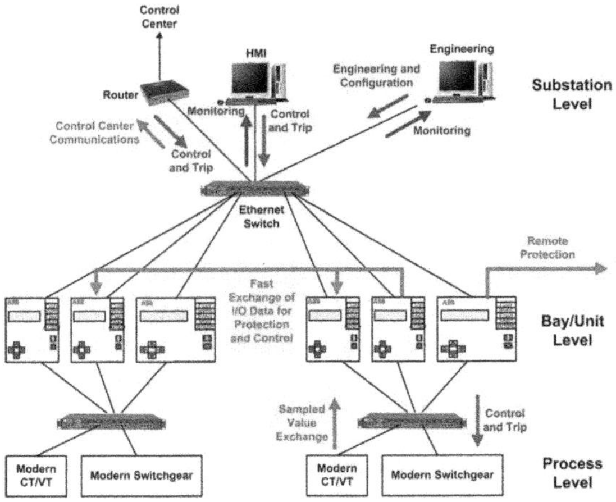


Figure 6-1- Substation automation architecture

### 6.2.1 Analysis of Cascade Architecture

A typical simple cascading (or bus) architecture is shown in Figure 6-2. As it is shown in the figure, each switch is connected to the adjacent switch via a communication channel. Depending on the tolerated delay and the network limit, the number of switches can be increased. This structure is economical and simple in configuration. However, the structure is not reliable enough for the critical application. As it is clear, if a connection between switches is lost, the communication to other IEDs may also be lost because of that failed connection. To increase the reliability of this structure, the redundant network can be applied. In this case, a single failure in the connection cannot cause the entire system to fail.

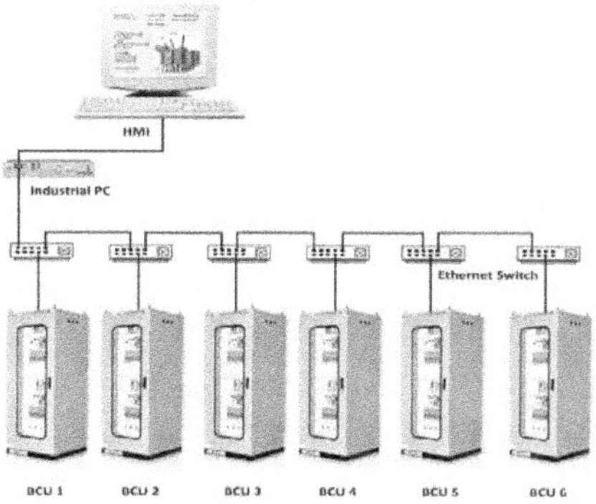


Figure 6-2- Cascade architecture

### 6.2.2 Analysis of Redundant Cascading Architecture

A redundant cascading architecture is illustrated in Figure 6-3. In this structure, the bay devices are connected to two separate cascade networks which provide a higher reliability than cascade structure.

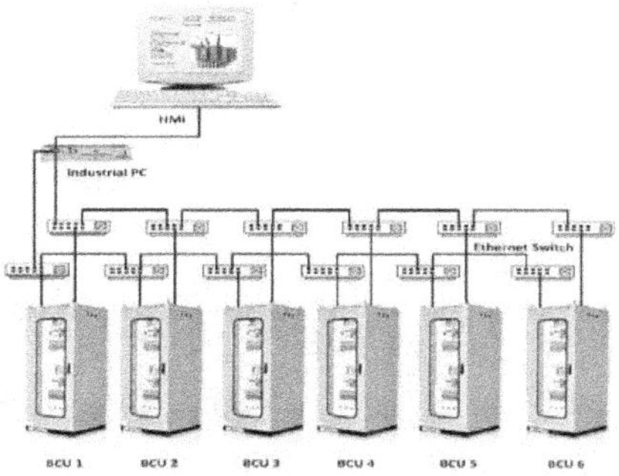


Figure 6-3- Redundant cascading architecture

### 6.2.3 Analysis of Ring Architecture

To provide higher accuracy, the cascade structure can be closed in a loop. This structure is known as the ring which is shown in Figure 6-4. As compared to redundant cascade, this topology is more economical and it has immunity to single network failure. However, the latency of critical application should be considered in this structure.

### 6.2.4 Analysis of Star Architecture

A typical star architecture is shown in Figure 6-5. As it is clear from the picture, ESW 1 is referred as the backbone of this structure and the other switches are connected to this switch. This architecture offers the least amount of latency. However, it has not sufficient redundancy due to the weakness of backbone switch.

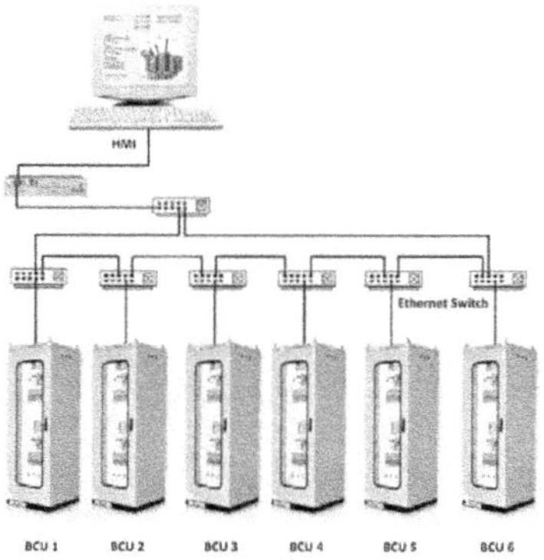


Figure 6-4- Ring architecture

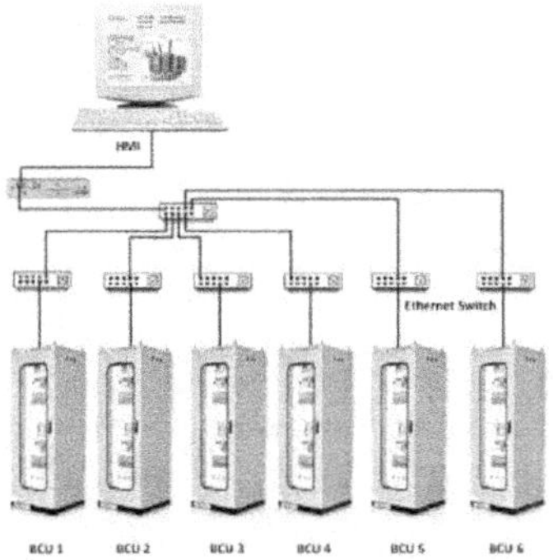


Figure 6-5- Star architecture

### 6.2.5 Analysis of Hybrid Architecture

To increase the reliability of star topology, the combination of star and ring are known as the hybrid can be applied which is shown in Figure 6-6. It provides higher reliability and is immune to many types of faults.

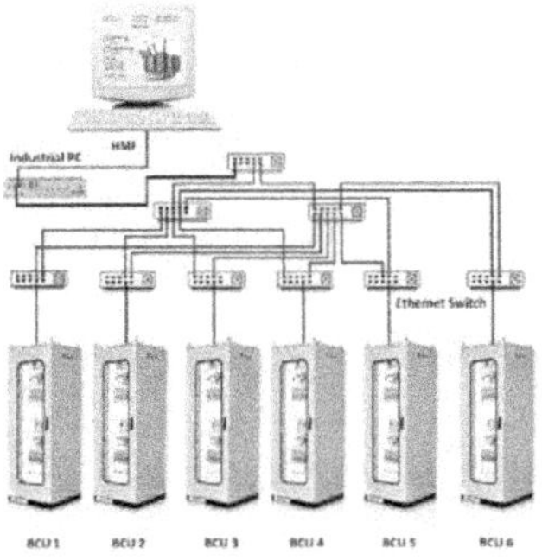


Figure 6-6- Hybrid architecture

## 6.3 Reliability Block Diagram (RBD) Equations

The first step in Reliability Block Diagram (RBD) technique to evaluate the reliability is to develop the RBDs for the system. The components which they are used together to perform a function should be modelled as series logic in RBD, and the redundant component should be modelled as parallel [127, 140, 141].

Following are the Formulae to calculate MTTF and MTTR which they can be applied to evaluate the reliability of a system;

$$R(t) = e^{-\lambda t} \quad \text{Eq. 6.1}$$

$$MTBF = \frac{1}{\lambda} \quad \text{Eq. 6.2}$$

Where $R$ is reliability, $MTBF$ is meat time between failures and $t$ is time.

In this chapter, we use block diagram to calculate the MTBFs. In this method, we should make a block diagram of series and parallel component. If the failure of a component fails the system, it should be modelled with series blocks (Figure 6-7). The reliability and MTBF in the series block can be calculated by Eq. 6.3 and Eq. 6.4.

Figure 6-7- Series block

$$R_S(t) = e^{-\lambda_1 .t} . e^{-\lambda_2 .t} = e^{-(\lambda_1 + \lambda_2).t} \quad \text{Eq. 6.3}$$

$$MTBF = \frac{1}{\lambda_1 + \lambda_2} = \frac{1}{MTBF_1} + \frac{1}{MTBF_2} \qquad \text{Eq. 6.4}$$

Where $R$ is reliability, $MTBF$ is meat time between failures, $MTBF_1$ is mean time between failures in the first component, $MTBF_2$ is mean time between failure in the second component and $t$ is time.

If a system with a failure of a component continues the process without any interruption, this component should be modelled by parallel block in the RBD which is shown in Figure 6-8. The reliability and MTBF in the parallel block can be calculated by Eq. 6.5 and Eq. 6.6.

Figure 6-8- Parallel block

$$R_P(t) = e^{-\lambda_1 . t} + e^{-\lambda_2 . t} - e^{-(\lambda_1 + \lambda_2) . t} \qquad \text{Eq. 6.5}$$

$$MTBF = MTBF_1 + MTBF_2 - \frac{MTBF_1 . MTBF_2}{MTBF_1 + MTBF_2} \qquad \text{Eq. 6.6}$$

Where R is reliability, MTBF is meat time between failures, $MTBF_1$ is mean time between failures in the first component, $MTBF_2$ is mean time between failure in the second component and t is time.

## 6.4 Reliability Block Diagram Modelling in a Substation

We evaluated the reliability of a simple substation with one incoming feeder and two outgoing feeders with one transformer bay and one bus-bur bay which is shown in Figure 6-9.

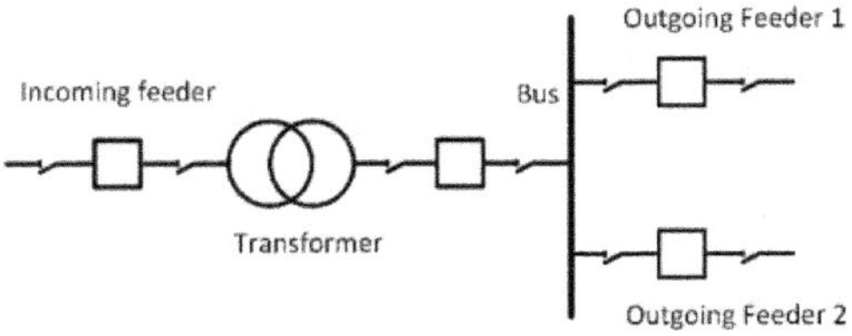


Figure 6-9- Simple substation for study

Table 6-1 summarizes the required units for each bay. And Table 6-2 summarizes the MTBF of each equipment [128, 142].

Table 6-1- Required units for each bay

| **Bay Name** | **Protection Unit** | **Control Unit** | **Measurement Unit** |
|---|---|---|---|
| Incoming Feeder | 2 | 1 | 1 |
| Transformer | 2 | 1 | 2 |
| Bus | 2 | 1 | 2 |
| Outgoing Feeder 1 | 2 | 1 | 1 |
| Outgoing Feeder 2 | 2 | 1 | 1 |

Table 6-2- MTBF of equipment

| equipment | MTBF (year) |
|---|---|
| Protection Unit (PU) | 150 |
| Control Unit (CU) | 150 |
| Measurement Unit (MU) | 150 |
| Network Switch (SW) | 50 |
| Synchronizing Unit (SYNC) | 150 |
| Connection Links (Link) | 150 |

### 6.4.1 RBD Analysis of Feeder Bay

Figure 6-10 represents the feeder bay is connected directly to a switch with the star topology. Figure 6-11 illustrates the reliability block diagram for a feeder bay. As there is a redundant protection unit in feeder bay, it should be modelled with the parallel diagram. Based on the definition of series and block diagram, the other components are modelled in series and parallel diagram. The MTBF for this bay can be calculated with Eq. 6.4 and Eq. 6.6 as follow:

$$\frac{1}{MTBF} = 0.056657 \quad \text{Eq. 6.7}$$

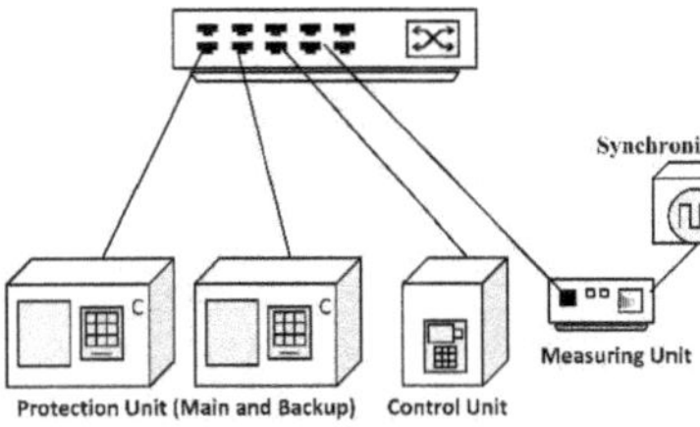


Figure 6-10- Feeder Bay

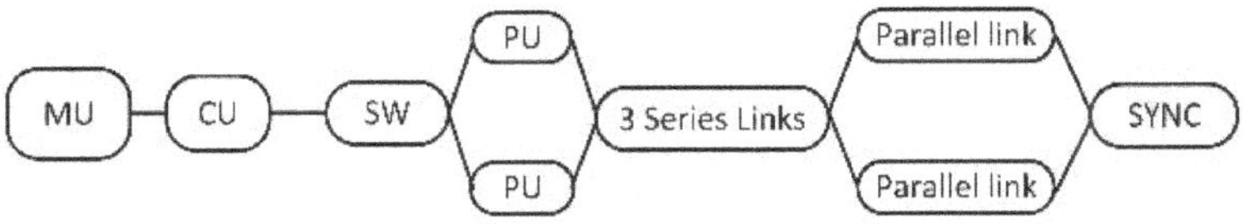


Figure 6-11- RBD for feeder bays

### 6.4.2 RBD Analysis of Transformer and Bus-Bar Bays

Figure 6-12 represents the transformer and bus-bar bays structures. Figure 6-13 illustrates the reliability block diagram for the transformer and bus-bar bays. As there are redundant protection units in these bays, they should be modelled with parallel diagram. Based on the definition of series and block diagram, the other components are modelled in series and parallel diagram. The MTBF for these bays can be calculated with Eq. 6.4 and Eq. 6.6 as follow:

$$\frac{1}{MTBF} = 0.07003 \qquad \text{Eq. 6.8}$$

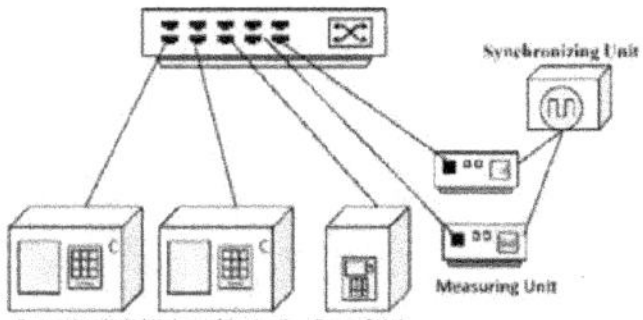


Figure 6-12- Transformer and bus-bar Bays

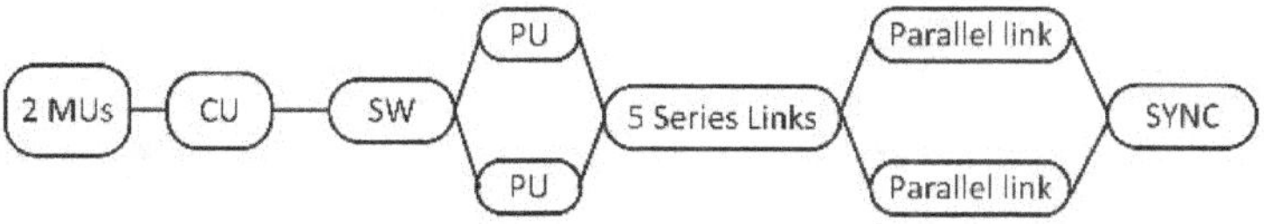


Figure 6-13- RBD for Transformer and bus-bar bays

## 6.4.3 Reliability Evaluation for SAS Architecture

This section evaluates the reliability of different structure for the substation which is discussed in the previous section. To evaluate the reliability, the MTBF for different topology is calculated.

### 6.4.3.1 RBD Analysis of Bus Topology

Figure 6-14 shows the bus or cascade topology for the substation. All elements in this structure should be modelled as the series block diagram. The MTBF for this topology is 6.67 based on the calculation by Eq. 6.4 and Eq. 6.6

### 6.4.3.2 RBD Analysis of Ring Topology

The next topology is the ring which is shown in Figure 6-15. It is clear that the reliability can be improved in this topology, as there is inherent backup communication link. In the other words, if there is the failure in a switch, the system will not fail. The MTBF for this topology is 7.69 based on the calculation by Eq. 6.4 and Eq. 6.6.

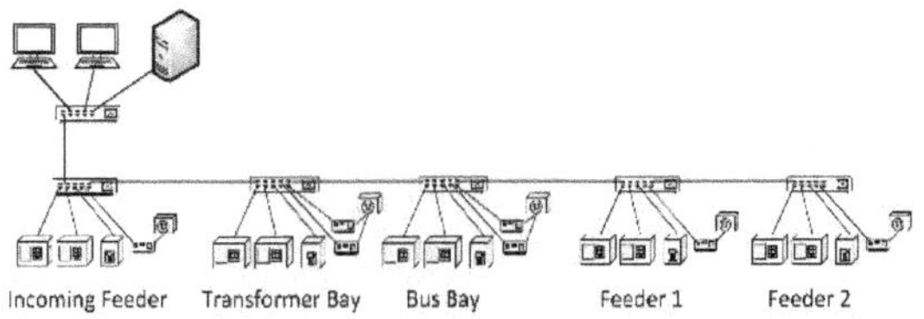


Figure 6-14- Bus topology

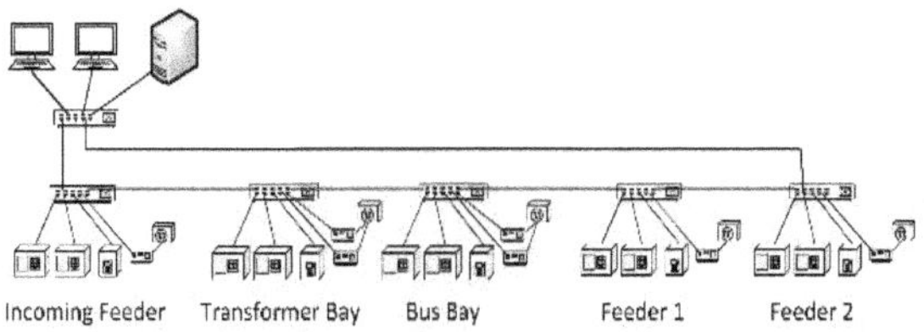


Figure 6-15- Ring topology

#### 6.4.3.3 RBD Analysis of Hybrid Topology

Figure 6-16 shows the hybrid topology. To improve the reliability, the combination of star and ring is utilized in this topology. The MTBF for this topology is 13.63 based on the calculation by Eq. 6.4 and Eq. 6.6.

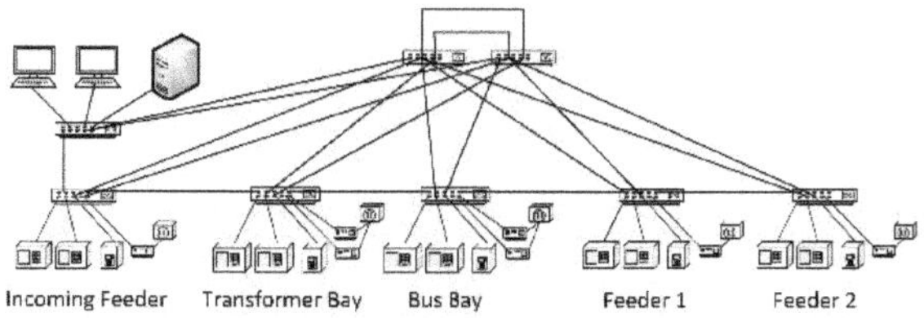


Figure 6-16- Hybrid Topology

## 6.5 Process Bus Topologies

Depending on reliability and data flow requirements, there are many possibilities for process bus communication. Figure 6-17 shows the four common alternatives for in process bus of transmission substation [143]. In the first alternative, each bay has its own process bus segment. A router is installed at the end of each segment to transmit the required data. In the second alternative, each bay segment covers more than one bay. In the third alternative, all devices are connected as a single communication bus. In the fourth alternative, the bus segments are set up to corresponding protection zones.

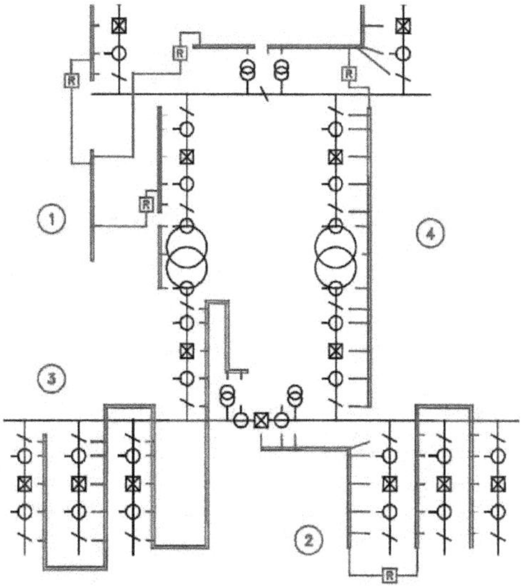


Figure 6-17- Process bus communication solutions [143]

As in IEC 61850, both process bus and station bus are Ethernet based, they can be integrated. However, in practical case, they are usually separated from each other. From reliability points of view, any of the topologies which are mentioned in the previous sections for station bus could be used for process bus too and it may have a different topology as compared to the station bus.

## 6.6 High Redundancy Protocols

Although the redundancy of communication is not required in IEC 61850, the customer is using it to increase the availability of communication network in SAS. The first edition of IEC 61850 does not include any selected technology for redundancy. However, the second edition of the standard refer to IEC 62439. IEC 62439 defines a set of Ethernet-based protocols providing high availability features to prevent loss of data due to single failure in the industrial networks [144]. Table 6-3 summarizes the properties of the different protocols based on the IEC 62439 standard [145].

Table 6-3- Properties of the different protocols based on the IEC 62439 [145]

| Protocol | Frame Loss | Topology | Recovery Time (ms) |
|---|---|---|---|
| MRP | Yes | Ring | 500 for 50 nodes |
| HSR | No | Ring | 0 |
| PRP | No | Mesh | 0 |
| BRP | Yes | Mesh | 4.8 for 500 nodes |
| DRP | Yes | Ring | 100 for 50 nodes |
| CRP | Yes | Mesh | 1000 for 512 |

Bumpless redundancy or seamless redundancy is a crucial protocol property for the critical application. Only High-availability Seamless Redundancy (HSR) and Parallel

Redundancy Protocol (PRP) can provide bumpless redundancy in case of a failure of an industrial network. Following subsections describe these two protocols in SAS networks.

### 6.6.1 High–availability Seamless Redundancy (HSR)

The basic concept behind HSR is that all devices are connected as a ring topology without switch. Figure 6-18 shows a typical HSR network where the source device publishes the "A" and "B" frames out of the two ports. A destination device receives these frames within a certain interval and it accepts the first frame and discards the second one. The interval time is negligible hence zero recovery time is achieved.

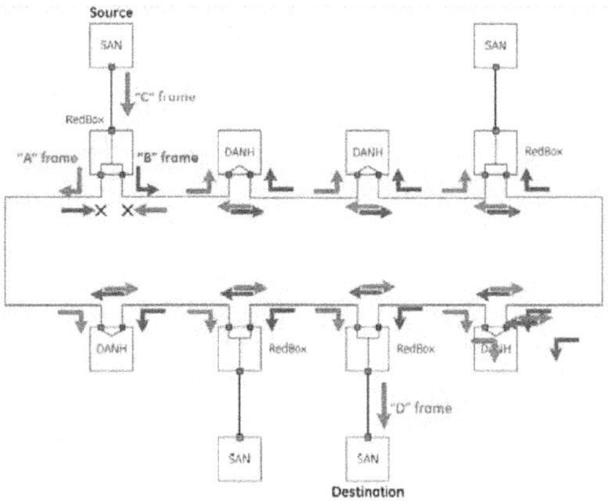


Figure 6-18- HSR concept

### 6.6.2 Parallel Redundancy Protocol (PRP)

The basic concept behind PRP is that a device is connected to two networks. These networks must be completely independent of each other. A device which is not supporting PRP, is considered a SAN which can be connected as a single node to any of the network. It can only communicate with other devices within the same network. However, with a RedBox, it is possible to attach a SAN to a PRP network. The RedBox mirrors the published frame in the source node and it creates the frame from the mirrored PRP to send to a SAN in the destination node. Figure 6-19 shows the general concept of PRP. The "C" Frame is published by device A which is mirrored as the PRP "A" frame and "B" frame to LAN A and LAN B correspondingly. The subscribed device is connected to the both LANs too. It accepts the first received frame and discards the second one. The resulting output is PRP "D" frame in the destination. As it is clear, if the LAN A fails, the "B" frame will be received by LAN B as normal.

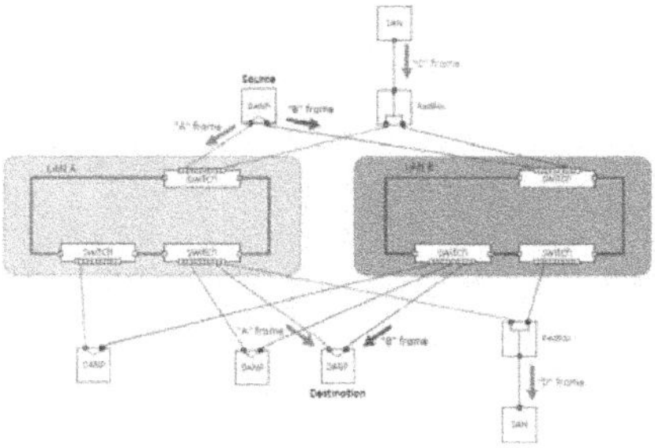


Figure 6-19- PRP concept

## 6.7 Implementation of the Process Bus in the lab

The demonstration and testing of different process bus communication topologies have been carried out on laboratory testbed developed at Substation Automation System (SAS) lab at Jamia Millia Islamia University, New Delhi India (Figure 6-20). This test bed comprises of industrial process bus equipment (AMU), secondary test kit (CMC 256-6), manageable Ethernet switches (Magnum 6KL), and Intelligent Electronic Devices (IEDs) such as (SEL-2407) as shown in Figure 6-20.

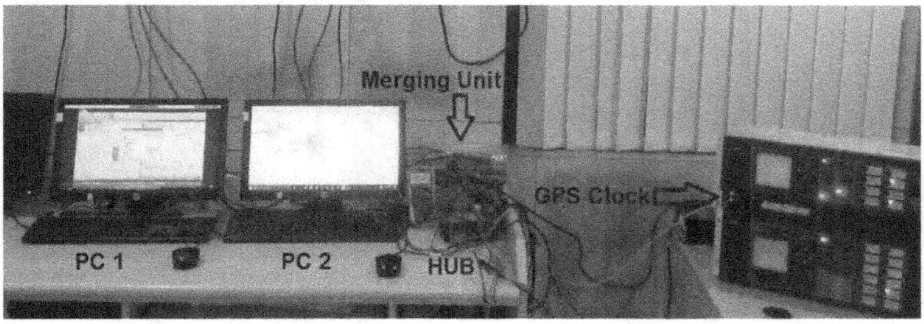


Figure 6-20- Lab setup at Jamia Millia Islamia University, New Delhi India

The MU combines the acquisition of measured voltage and current from conventional instrument transformer into IEC 61850 Sample Values (SVs). These SVs will be published according to the SV control block for protection and measurements. Figure 6-21 shows the hardware configuration of merging unit in the lab.

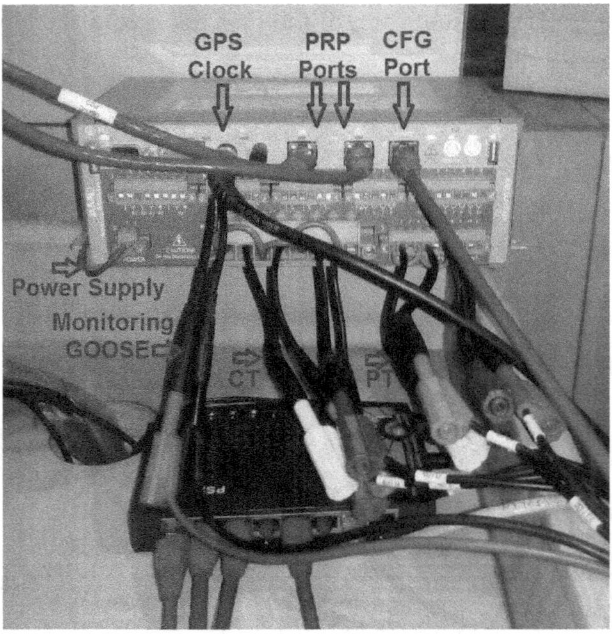


Figure 6-21- Merging Unit (MU) hardware configuration

Several experiments based on the lab setup have been carried out to evaluate different topologies in the process bus. Based on the observation, the most practical and useful topologies for the process bus in the lab are star, ring, and the redundant them.

Figure 6-22 shows star topology where the Hub is the backbone switch in this setup. Three phase signals are injected to the MU, the associated GOOSE and SVs messages are subscribed in the PC1 and PC2. Figure 6-23 shows the subscribed GOOSE messages in PC1 where the retransmission time is 2 sec. This architecture offers the least amount of latency. However, it does not have sufficient redundancy due to the weakness of backbone switch.

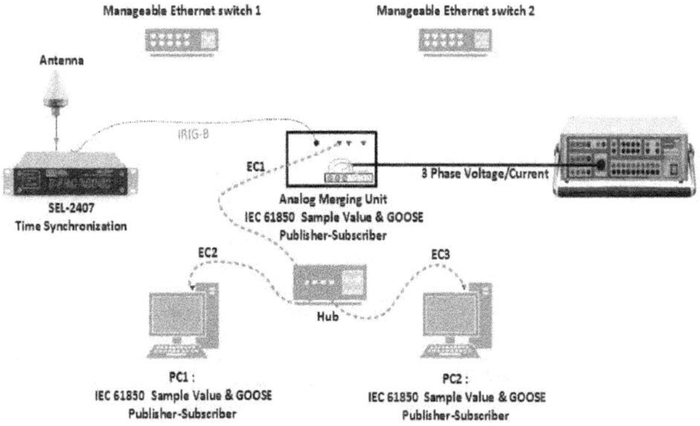


Figure 6-22- Star topology

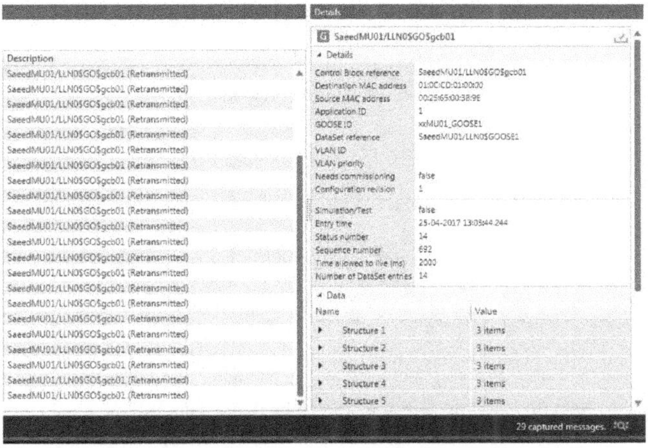


Figure 6-23- Captured GOOSE in star topology

Figure 6-24 shows star-ring topology where MU is connected to ESW 2 with EC1, PC 1 is connected ESW 2 with EC2, PC2 is connected to Hub with EC3. Also, ESW 2 is connected to ESW 1 with EC4, ESW 2 is connected to Hub with EC5, and Hub is connected to ESW 2 with EC6. This topology is economical and it has immunity to single network failure. However, latency of critical application should be considered in this structure.

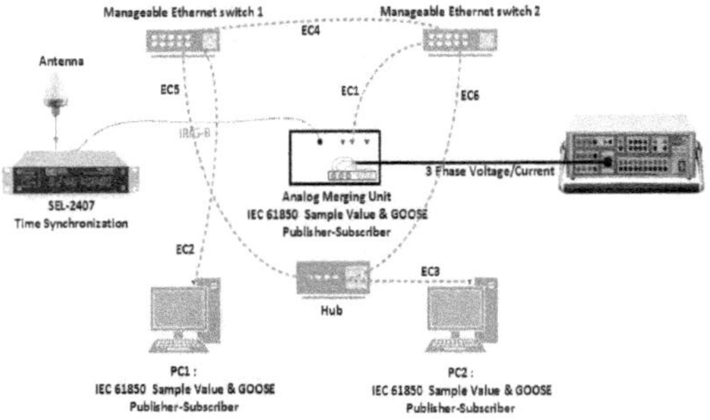


Figure 6-24- Star- Ring topology

The MU has two redundant ports and it supports PRP protocol. Figure 6-25 shows the installed PRP topology where port 1 in MU is connected to ESW 2 with EC2, and port 1 of the MU is connected to ESW 1 with EC 1. Pre-fault signals are injected to MU by the CMC 256-6 and the associated GOOSE and SV messages are subscribed in PC1 and PC2. Figure 6-26 shows the subscribed GOOSE in PC1. In this topology, it is observed that after each re-transition signal, one GOOSE message is received. Also, it is shown in Figure 6-26 that after removing a redundant cable, the network acts similar to a simple network without transmitting GOOSE after each retransmission.

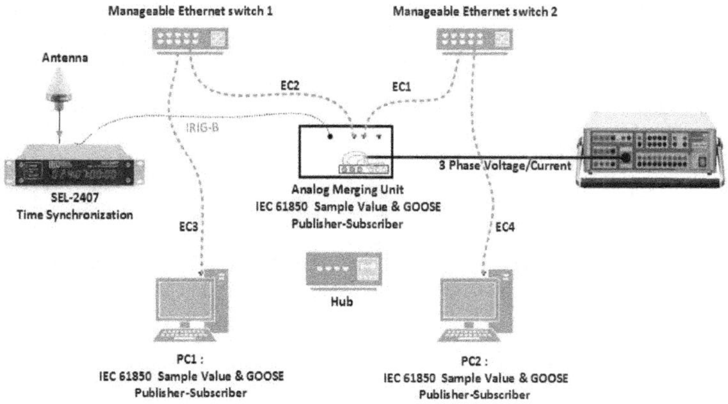


Figure 6-25- RPR topology

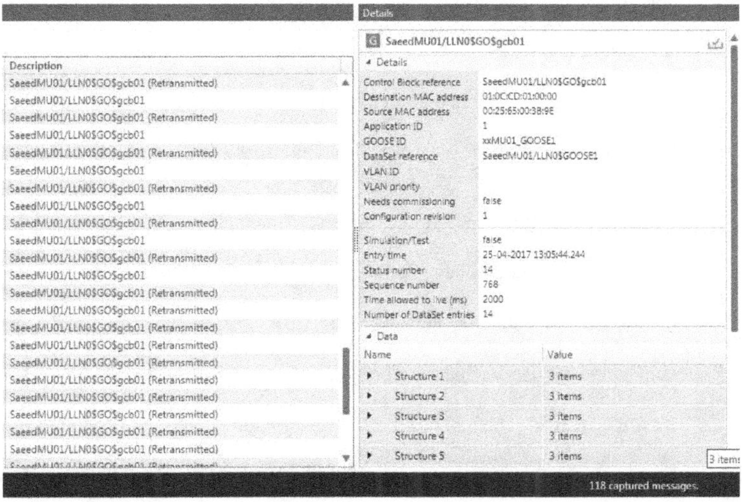


Figure 6-26- Captured GOOSE in redundant topology

There are many advantages and disadvantages of using PRP network. The first benefit is that due to use of mirroring frames, PRP achieves no interruption of communication. Another benefit is that the PRP networks can work in any topology. And the most important advantage is that it is possible to apply VLANs, message priority, and MAC address filtering for GOOSE and SVs [146]. The disadvantage from a technical point of view is that the two PRP networks cannot be connected together in any way.

## 6.8 Conclusion

This chapter presents the network topologies for a substation automation system applying IEC 61850; also it presents an approach for reliability assessments of various substation automation systems. The reliability block diagram has been applied in various substation automation architectures. The results show that both the ring and hybrid topologies are the most reliable architecture in comparison to cascade topology.

The use of the IEC 62439 is one of the fundamental parts of the process bus network. HSR and PRP protocols have been introduced in IEC 62439 to address the requirement of the reliable industrial network. The basic concept involved in the protocol is duplicating the data exchange to provide zero recovery time in case of single failure of a switch or link. This chapter presented different process bus topologies using the benefits of PRP protocol. A lab setup was developed to implement the process bus network communication and to evaluate the different network topologies. It was observed that PRP was the most robust of the topologies for process bus where there are two separate networks and there is zero recovery time in case of a failure in one of the network, it can be readily integrated with other topologies. However, the implementation doubles the costs for network component such as switches and cabling.

The next chapter discusses the outcome of the entire research work in the form of conclusion. It also provides the future scope for research work in this area.

# Chapter 7 Conclusion and Future Work

## 7.1 Summary

Communication technology has developed very fast and smart devices like IEDs are using IEC 61850 for communication. With such a technology, new distributed functions based communication can be investigated. However, there is less research on the distributed functions. This research work has been investigated to transmission line protection functions. New schema based distributed measured values have been developed in fault location and line differential protection functions. The source code of IEC 61850 has been explored and customized for applying of IEC 61850 GOOSE based interlocking and transmission line inter-trip schemes. Also, a detailed communication profile has been described to show how to configure IEC 61850 GOOSE in a lab environment which is beneficial for engineers in industry and academicians for implementing IEC 61850 in such advanced IEDs.

The following conclusions have been achieved at after extensive investigation, which is further substantiated by the research work presented in this thesis:

Impedance-based fault location algorithms are used to estimate the location of faults on a transmission line. These methods are commonly used by utilities because of simplicity. However, there is no specific test module in the relay test kits to evaluate this important function. This work proposed a method to test the function in protection relays. To test the scheme, laboratory setup is developed. Several faults in a transmission line are simulated in PSCAD and associated fault data are injected by CMC 256-6 to a protection relay IED (SEL 421). The process of testing is presented with details and the experimental results are analysed.

Currently, TW-based fault location is standalone device. A new technique is developed to integrate TW-based fault location with existing protection relay. Despite analysing three phase system in the most proposed algorithm by other researchers, the technique in this research applied single input for the algorithm. The input used to detect the fault in a line which can be monitored with high sampling frequency in a protection relay.

Differential and distance with back up zones have been the main protection functions in transmission lines. The performance of these functions has been compared with several faults with different features. Simulation results have shown the differential protection is more fast and sensitive in resistive faults. However, in practice, the differential relay doesn't have satisfactory performance due to charging current of long transmission lines. This work proposed to divide a long transmission line to many short lines. And it is proposed to distributed fiber optic current sensors in a line. The measured values of these sensors should be sent to substations throughout communication channels. A new current differential protection is developed for the proposed schema. Simulation results show that proposed differential function is more sensitive to detect a resistive fault than existing differential function.

In order to increase the fault clearing time, IEC 61850 GOOSE based distance protection inter trip scheme is presented in this thesis. IEC 61850 Ed2 standardize communication between substations which is enabled new protection schemes based on the distributed values in the substations. This thesis discussed adopting IEC 61850 GOOSE based accelerated distance protection. The tunnelling concept has been presented to send and receive GOOSE messages between substations. A system in loop is developed to analysis the delay of GOOSE messages for different scenarios. The performance evaluation carried out for different types of WAN links and it has been concluded that links with lower bandwidths are not suitable for long distances as they may result in the high end to end delay.

A laboratory setup has been developed to test and implement IEC 61850 GOOSE functions. Bay interlocking function was simulated in PSCAD and practical implementation of interlocking based IEC 61850 GOOSE has been implemented in the lab. Replacing hard wire with IEC 61850 GOOSE has many advantages such as cost saving in installation, test and maintenance, flexibility for subsequent modification, safety by replacing electrical circuit with communication, and live monitoring of signals. Although many scholars published many research papers in the area, practical installation of this technique has many obstacles due to lack of documents and tutorials.

A laboratory setup has been developed to test the performance of IEC 61850 process bus in different topologies. The PRP protocol was the most reliable method in process bus, and it could integrate with other topologies. However, the cost of implementation, especially for network component, is higher than simple one.

## 7.2 Future Research Work

The scope of future research work is as follow:

The proposed fault location scheme is tested in the un-transposed overhead line. This scheme can be evaluated for the other types of transmission lines such as:

- Underground transmission line.
- Double circuit transmission line.
- Overhead line and cable composite network
- Series-compensated line
- Transposed line

Other distributed operation and protection functions can be evaluated based on IEC 61850.

The analog signals are replaced by an Ethernet message that contains samples from a merging unit connected to CTs/PTs. In this thesis fault location based on the process bus data was presented. New protection scheme based on the IEC 61850 process bus can be developed.

Techno-economic evaluation of the proposed fault location scheme can be carried out.

Several IEC 61850 process bus topologies were evaluated based on the experimental results. The performance of different applications and functions can be evaluated in these topologies.

www.ingramcontent.com/pod-product-compliance
Lightning Source LLC
LaVergne TN
LVHW010726061126
851857LV00014B/1366
* 9 7 8 1 0 8 8 1 6 2 3 8 5 *